DYNAMICS OF PHYSICAL CIRCUITS AND SYSTEMS

**⌄ MATRIX SERIES
IN CIRCUITS AND SYSTEMS**

Andrew P. Sage, *Editor*

DYNAMICS OF PHYSICAL CIRCUITS AND SYSTEMS

JAMES F. LINDSAY

Associate Professor
Department of Electrical Engineering
Concordia University

SILAS KATZ

Associate Professor
Department of Mechanical Engineering
Concordia University

MATRIX PUBLISHERS, INC., Champaign, Illinois • Portland, Oregon

© COPYRIGHT, MATRIX PUBLISHERS, INC.   1978

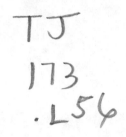

10 9 8 7 6 5 4 3 2 1

Library of Congress catalog card number: 78-53838

Matrix Publishers, Inc.
Champaign, IL 61820

ISBN: 0-916460-21-5

This book was prepared by typewriter composition.
Illustrations by Scientific Illustrators.
In-house Editor was Merl K. Miller.

CONTENTS

BASIC NETWORK MODELS **4**

RESPONSE OF FIRST ORDER CIRCUITS **5**

RESPONSE OF SECOND ORDER CIRCUITS **6**

GENERALIZED IMPEDANCES AND SYSTEM FUNCTIONS **7**

SINUSOIDAL RESPONSE 8

MUTUALLY COUPLED COMPONENTS 9

ANSWERS TO SELECTED PROBLEMS

BIBLIOGRAPHY

INDEX

PREFACE

To many of the students entering engineering or technology programs this is an instantaneous world. For them
transients either do not exist or are not very important.
They proceed from a wish to the initiation of an action
and then the desired result occurs zero seconds later.
The neglect of transients is not restricted to students
alone. It is common to all of us, occasionally. For
example, a package is to be moved from one place to another
and it gets there - somehow. The velocity during transit
usually does not concern us. Of course, if the package
contained a block of ice and it was to be delivered across
town on a hot day, we might give more thought to the transit
time. Similarly, when a television is switched on, we pay
little attention to the warm-up time. Has anyone ever seen
anybody measuring the duration of the warm-up? Yet if
important news was about to be broadcast and we feared
missing it, we would notice the warm-up period and be
irritated by it.

Since so much of our previous training and environment
emphasize instantaneous action (i.e., quick food, over-
night success) we feel that it is necessary to call

attention to dynamic effects as early as possible in a
student's education. Accordingly, this text considers
transients and is designed for use in the first or second
year of an engineering or technology program. In this way,
students are made aware at the beginning of their program
that there are no instantaneous processes in real life and
that sometimes processes require a long time to complete.
The precise dynamics of a process depend on geometric and
material properties which students will normally encounter
in subsequent years when they study fluid mechanics, elec-
tronics, heat transfer and vibrations. However, from a
few basic ideas presented here students will be able to
estimate and appreciate the dynamic performance of certain
processes before they are exposed to the more detailed
studies.

Our purpose in writing this text, therefore, is to present system dynamics in the simplest possible way. We
believe that this can be accomplished best through the use
of circuits and circuit components. Most students have
been exposed briefly to only electrical circuits in their
high school physics courses. Thus the circuit concept
offers an excellent framework with which to bridge the gap
from static to dynamic analysis. More important perhaps
is that circuit theory lends itself very well to analogy.
Through the use of circuit methods we may show that it is
possible to treat problems in fluid, thermal, and mechanical
systems as well as in electrical systems. For example, the
similarity between charging a capacitor and filling a bucket
of water is made obvious by comparison of the mathematical
descriptions of these processes. Of course, we might have
recognized this similarity intuitively. However, we find
that these processes are also analogous to the change in
velocity of a mass when a force is applied. This is not
quite so obvious. Thus the circuit concept proves extremely
useful in reducing physical situations in different physical
media to a common form from which a common type of solution
may be obtained.

We have tried to present the material in a logical order.
However, it seemed that there were always several things
that had to be introduced simultaneously. Chapter I contains some basic ideas on system response and attempts to
introduce the technical concepts that follow. In Chapter II
we deal with time functions. Although the emphasis is on
the specification of certain functions as system inputs
the purpose is rather broader. It is to make the student
feel comfortable in the mathematical formulations and
operations with time dependent signals. Chapter III
describes the circuit components in the different physical
media and stresses the analogy between them. It also re-
inforces the concepts of time dependent signals by applying these signals to the components. The combination of
components into simple circuits and the basic circuit laws
are introduced in Chapter IV. The initial and final properties of storage elements and the effect of initial
storage are also presented here. The reason for this is
that these properties are best described by referring to
their effects in circuits. Chapters V and VI present the
solution of the equilibrium circuit equations for first
and second order circuits respectively. The generalized
impedance concept is reserved for Chapter VII since it
facilitates the formulation of equations in more complex
circuits. This material is presented without recourse to
Laplace transforms but is based rather on the concept of
exponential inputs. This has the advantage that students
work in the time domain and thereby develop a sound appreciation of the dynamic behavior of physical situations.
Chapter VIII considers the frequency response of circuits
and this follows quite properly from the impedance concepts.
Finally mutually coupled devices are considered in Chapter
IX. This provides an introduction to two-terminal pair
circuits primarily in the form of the ideal transformer
and its analogs.

We acknowledge the strong encouragement of Professor
J.C. Callaghan, President of Nova Scotia Technical College

who was Dean of Engineering at the time we were writing the
text. We are also indebted to Professors M.N.S. Swamy,
Dean of Engineering, and T.S. Sankar, Chairman of Mechanical Engineering. We appreciate the contributions of
Dr. F. Blader (now at the National Research Council of
Canada), Dr. S. Tsang (now at Lakehead University),
Mr. C. Trueman and Mr. O. Ahmad who have taught this
material during its development and provided many helpful
suggestions. We thank Mr. R. Gallant for his enthusiasm
and diligence in preparing many of the problems used in
the text. He also gave us a better understanding of the
difficulties that a student might have with the material.
In addition we are grateful to Dr. A. Sage, University of
Virginia who reviewed the manuscript and provided us with
many valuable comments.

We thank Miss M. Stredder who typed the manuscript while
it was in the form of class notes. Finally we are especially indebted to Miss L. Mikhail who typed the final copy.

J. F. Lindsay
S. Katz

Montreal, Canada
March 1978

CHAPTER 1

INTRODUCTION

1.1 THE CIRCUIT AS A SYSTEM

Most engineering endeavors are concerned with physical
devices or groups of devices for which the word "system"
is generally applied. By definition, a system is a set of
things or parts forming a whole and serving a common purpose.
The components or devices that combine to form a physical
system are based on phenomena that engineering students
normally encounter in a first course in physics. Our
purpose here is to introduce methods for describing and
analyzing physical systems.

The concept of a physical system is very broad. We attempt
to be more specific by restricting our consideration to

physical circuits. Basically a circuit is a closed path.
However, we use the term "circuit" to indicate a special
type of system in which the components represent only one
physical area. Thus we will deal with electrical circuits,
mechanical circuits, fluid circuits, or thermal circuits.
The advantage of the circuit approach is that the techniques
used for the analysis of electric circuits are highly
developed. These same techniques may be applied by analogy
to problems in mechanical, fluid, and thermal circuits. In
this way, the similarities between problems, in different
physical media, are made obvious. As a result, solutions
in one physical medium are transferable, by analogy, to the
other media.

   The description of a circuit may be qualitative or
quantitative. If the latter, there is often the necessity
for taking measurements, either to provide basic data or
to confirm that a design does what it is supposed to do.
The process of describing a physical situation is often
referred to as "modelling". To determine a suitable model
requires the ability to observe the physical situation and
to describe it first in qualitative terms. For example,
when an automobile passes over a bump in the road there is
an up and down motion which takes place immediately after-
wards. The manufacturer wants to know before building a
million cars whether the motion will be small or large. It
is necessary, therefore, to think in terms of a model that
can give quantitative information on the amplitude of the
motion. To obtain such a model, however, requires a more
detailed qualitative description of the physical phenomena
involved. In this case it requires an understanding of the
main parts of the suspension and the ability to observe
exactly what happens at the point of contact between the
wheels and the road. It is only after the qualitative
description is complete that the quantitative description
may be obtained.

   There is, of course, a temptation to believe that once
the quantitative description has been obtained (i.e., the

formula), the problem is solved. The danger in this atti-
tude is that physical situations may change. Springs may
break or shock absorbers may wear out. The only sure defense
against this is to observe continually the physical situation.

1.2 THE BASIC COMPONENTS

One of our primary objectives is the determination of
suitable models for physical situations. Specifically we
want mathematical models based on established natural laws.
However we limit our consideration to physical situations
where the mathematical modelling can be accommodated by
means of linear constant coefficient equations. Mathematical
models will be developed only for the most basic physical
components. More complicated situations can then be described with combinations of basic components.

For example, let us consider one of the basic fluid
system components - the fluid line. Figure 1.1a shows a
pump supplying fluid to a line. How are we to describe
this component mathematically? To answer this question we
must first decide what we wish to know about the passage
of the fluid in the line. One thing we want to know is the
rate at which liquid is passing through the line. We call
this the flow and for a fluid system it is designated as q.
Another thing, we might need to know is the size of the
pump required to produce the flow. Instead of physical size,
the capability of a pump is often indicated by the pressure,
p, it can develop. If we were able to relate the pressure,
p, to the flow, q, we would have a type of mathematical
model of the line. But can we do this? We cannot because
the flow through the line depends also on the pressure, p_A,
at the end of the line. If p_A were equal to p there would
be no flow through the line. Thus our mathematical model
must relate the flow, q, to the pressure difference, $p - p_a$.
These quantities, flow and pressure difference, are the
signal variables for fluid systems. Flow is the through

variable (T.V.) and pressure difference is the across
variable difference (A.V.). To determine the mathematical
model of the line we could perform some experiments on the
line in which the pressure difference was changed and the
resulting flow measured. Figure 1.1b shows some typical
results for three data points (indicated by x). As pressure
difference increases, the flow increases. The mathematical
model, of course, must hold for an infinite number of
pressure difference - flow combinations - not just the
three data points. To expand the data so that it will
prove useful for the other cases we use the data to draw
a straight line between the points (shown dashed on Figure
1.1b). This dashed line is based on our limited observations
and is assumed to represent the fluid component at values of
the signal variables where we have made no measurements. The
mathematical description of the dashed line is our model of
the fluid component.

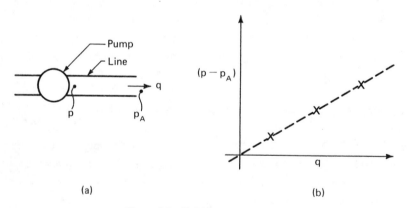

(a)   (b)

Figure 1.1 Fluid line component

It is possible to describe most of the basic physical
situations by means of two-terminal components, but there
are situations which require a model having two pairs of
terminals. The common two-terminal-pair components are
described in Chapter IX. However, at this stage we consider

only the two terminal components. The general two terminal
component is shown schematically in Figure 1.2. The across
variable difference represents a difference in a physical
condition on either side of the component and the through
variable represents the quantity transmitted through the
component. The value of the two terminal representation
is that it permits components to be combined together into
circuits that can be analyzed by a well developed circuit
theory. In this way we may use a common method of analysis
for all physical circuits.

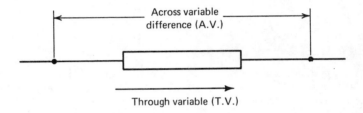

Across variable
difference (A.V.)

Through variable (T.V.)

Figure 1.2 Two terminal component

In general, there are three basic components in each of
the physical media:
  (a) one dissipative component in which energy is lost
      and
  (b) two storage components in which energy is retained.
Figure 1.3 shows the circuit symbol for each component
performance. The symbol for the dissipative component,
Figure 1.3a, usually represents electrical resistance.
However, we use this symbol to represent a component in
any of the media in which the across variable difference
and through variable are directly proportional to each
other. This is the situation that fits into our observation
of the fluid line. Thus the fluid line may be considered as
a fluid resistance. In a similar way the storage element
symbols shown in Figure 1.3b and c, usually indicate electrical inductance and capacitance. For our purposes we use

these same symbols for components in other media that produce
the same relationship between the through and across variables. We shall find, for example, that mechanical springs
produce the same signal variable relation as an inductance
and that a mass produces a capacitive type relation. These
components are discussed, in detail, in Chapter III. For
the present, however, it will be sufficient to observe that
the relations between the variables are either proportional,
differential, or integral. The differential and integral
components perform time dependent operations on the signal
variables. These mathematical operations are discussed
fully in Chapter II.

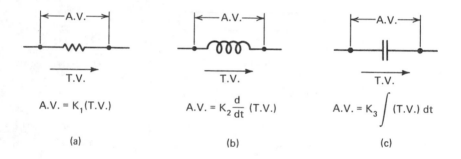

$$A.V. = K_1(T.V.) \qquad A.V. = K_2\frac{d}{dt}(T.V.) \qquad A.V. = K_3\int(T.V.)\,dt$$

(a)   (b)   (c)

Figure 1.3   Basic components

1.3 INPUT AND OUTPUT SIGNALS

Initially the circuits that we deal with consist of
particular combinations of two terminal components. These
combinations act together to produce a more complicated
mathematical function than is produced by a single component. However, in Chapter IX we introduce mutually coupled
components which have more than two terminals. A typical
circuit with two terminal components is shown in Figure 1.4.
The input represents a signal variable (either across or
through type) that is applied to the circuit. The input

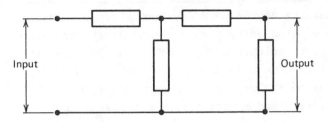

Figure 1.4 Typical circuit

is, therefore, basically a signal. The circuit operates on the input signal according to the circuit mathematical function and produces an output signal. The output signal is sometimes called the response of the circuit. It may also be either of the across or through type.

In normal use, a system or circuit produces some results, hopefully those desired by the person using the device. The usual term used for such results is "output". That is, the "output" of an automobile could be its velocity. However, in order to get an output, there must be some process of giving information or energy (or both) to the system. This is the "input". Inputs to an automobile may include the energy stored in the gasoline, position of the steering wheel, road surface, accelerator position, and brake pedal position. These concepts are usually shown graphically by means of a block diagram such as shown in Figure 1.5.

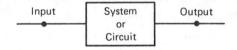

Figure 1.5 Block diagram representation of circuit

For an automobile, if efficiency is of no concern, there is no need to consider the output power as one of the outputs nor is there any need to consider the chemical energy of the gasoline as one of the inputs. In this case, the output torque of the engine would be an input to the automobile transmission system.

The meaning of the word "system" therefore includes the concept of a boundary within which all the parts constitute the system. In this example of the automobile, it must be noted that it is only the energy conversion process which would not be included in the system; the mass of the engine, both its case and moving parts, would still be included in the system. To a certain extent, the choice of the boundary for a particular problem is an art, the mastery of which is one of the attributes of the best engineers.

The terms "input", "output" and "system" may therefore seem to be rather nebulous. Nevertheless it is essential to be fully aware of what initiates the action in a system (input) and what constitutes the results (output). Usually there is little difficulty in determining the inputs and outputs of a system if it is possible to give a clear qualitative description of it.

The prediction of the behaviour of a physical system involves determination of the response (output) to certain hypothetical stimuli (inputs). Although many systems eventually operate under steady conditions (e.g. constant velocity for an automobile) there is the problem of getting them into such a steady state. Systems generally pass through a dynamic or transient state in which numerical values for their outputs are changing. Sometimes this dynamic state is the critical attribute in determining the acceptability of the performance, and therefore it is the dynamic response which must be determined. Since the numerical values of the outputs change with time, the simplest way to describe dynamic response is by means of a mathematical function of time. In general, the inputs also change with time and are described by mathematical time functions. However, there is the inherent understanding that the input function can be specified. It is the cause. The output function, on the other hand, is the effect and it depends on both the input and the circuit.

Figure 1.6 shows two typical input functions, the step input (Figure 1.6a) and the sine wave input (Figure 1.6b).

In an actual system the input may not be known in advance.
Some drivers step down hard on the gas while others drive
sensibly. However, the input functions shown are really
test inputs. They enable us to compare the performance of
systems and circuits. The output that results from a step
input is termed the step response. The output from a sine
wave input is the sinusoidal response. It consists of a
transient portion and a sinusoidal "steady state" portion.
The sine wave portion of the output has the same frequency
but, in general, a different amplitude and phase than the
sine wave input. The manner in which the amplitude and
phase of the outputs for input sine wave changes over a
range of different frequencies is called the frequency
response.

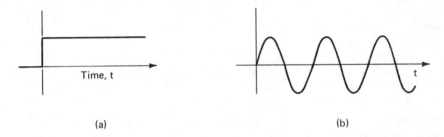

(a)   (b)

Figure 1.6 Typical inputs

An examination of the physical nature of some typical
responses should prove useful. Consider a simple pendulum
having a mass at the end of a string, Figure 1.7a. If a
constant force is applied periodically, the pendulum can
be made to swing with a constant amplitude. If the force
is no longer applied, the pendulum will continue to swing,
although its amplitude will decrease due to the air resistance (Figure 1.7b). The main point to note is that there
is motion (output) even when there is no applied force
(input). The form this motion takes is a property of the
pendulum and is called the "natural response" or "force-
free response". The force is applied periodically to take

advantage of this fundamental property of the system, the
difference between the natural response and the actual
response being the component of the response directly
attributable to the applied force. It is usually called
the "forced response".

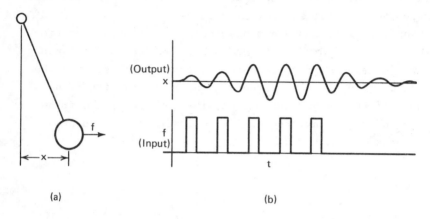

(a)   (b)

Figure 1.7 Response of pendulum

An example in which the forced response is more easily seen
is that of a bicycle being pedalled along a horizontal road
on a calm day. If the rider applies a constant force to the
pedals, the speed will increase until a steady value is
reached (Figure 1.8). This speed is directly related to the
applied force and is the forced response of the system.
After the rider stops pedalling, the bicycle still moves,
although at a decreasing speed. Again there is a response
without any stimulus. This is the natural response of the
system which can also be identified as being the difference
between the forced response and the actual response during
acceleration. Other examples in which the forced and natural
responses may readily be identified by a similar argument
are the heating and cooling of an oven and the switching on
and off of a large fan.

The mathematical model of any system must be capable of
giving solutions which adequately describe the physical

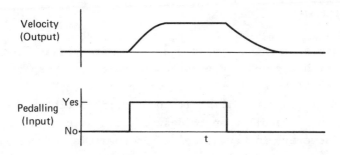

Figure 1.8 Response of bicycle

situations considered above. The model of a dynamic system
often is put into the form of one or more differential
equations, the individual terms of which describe each
individual or elemental effect within the system. The
solution of a differential equation has two parts: the
particular-integral solution and the complementary solution.
The particular-integral solution is identified with the
forced response and the complementary solution with the
natural response. Indeed, in an engineering context it is
usual to refer to these solutions by their physical descriptions, i.e. the forced and natural responses. For linear
systems (i.e. systems in which each elemental part has the
property that its response is directly proportional to the
input or its time derivative or time integral) the two component solutions are quite independent of each other. The
forced response (both physically and mathematically) has
exactly the same form as the input or driving function. The
natural response has a form which is completely independent
of the driving function although the magnitude does depend
on the input and also on the initial stored energy. Unfortunately, non-linear systems do not have this simplifying
property and solutions are significantly more difficult and
outside the scope of this text. The solution of linear
differential equations that are often encountered in physical
circuits is considered in Chapters V and VI.

1.4 MODELLING

The elemental effects within a system are physical phenomena which may be concentrated at one point in space, but
more generally are distributed over a region. As might be
expected, it is more difficult to describe a distributed
phenomenon in detail than to describe the situation at a
point. However, in many cases it is possible to consider
a limited number of points and to produce a model which
considers the action to take place at these discrete points,
yet still gives a sufficiently accurate description of the
interaction with the remainder of the system. For example,
a coil of wire produces a magnetic field throughout its
length; the conductor also has resistance which is distributed similarly. Despite the fact that in a real coil
these two effects exist simultaneously, it is in a great
many cases quite acceptable to consider the resistive effect
entirely separately from the magnetic inductive effect, and
the usual model consists of two elements: one is a conductor
which has resistance but produces no magnetic field and the
other is a conductor which produces a magnetic field but has
no resistance. This situation is analogous to the performance of a fluid line where we may also distinguish two
effects. One effect is the frictional dissipation that we
have considered in Figure 1.1. Another effect, that we had
neglected, is the one due to the inertia of the fluid. As a
result the fluid line may also be considered as two components in series.

Another example is that of a spring, (Figure 1.9). The
material of which the spring is made inevitably has mass
and therefore there must be an inertia force involved in
any change in velocity of the spring (Figure 1.9a). However,
this mass may be negligible in comparison with the mass of
the object to which the spring is connected. In these
circumstances, the spring is considered to exhibit only the
characteristic of stiffness. If this is not sufficiently
accurate, part of the mass of the spring is sometimes

considered as being situated at one end (Figure 1.9b). That
is the actual spring is represented by an ideal spring and
a mass connected to one end. It is fairly common to
describe practical situations which may reasonably be
represented by an ideal spring as "spring-like". Similarly,
if only the mass of an object is significant it is "mass-
like".

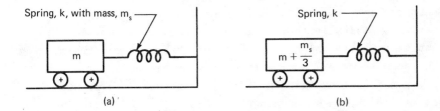

Figure 1.9 Modelling a spring having significant mass

This process of lumping each phenomenon into a simple
elemental situation produces a "lumped" model as opposed to
a "distributed" model. Since distributed models usually
require the solution of sets of partial differential equations and lumped models involve ordinary differential equations, there is a definite preference to use lumped models
whenever possible. Inevitably there are doubtful situations
in which it is not clear that the conditions are such that a
lumped model is valid. In such cases it may be necessary
to obtain both lumped and distributed models and to compare
solutions. Alternatively the lumped model may be compared
with experimental data. Either approach is necessary to
provide the experience which is needed for the development
of reliable models.

1.5 SIGNAL VARIABLES

We have touched on the signal variables briefly in Section
1.2 and they will be elaborated upon in Chapter III. However, they are so basic to physical circuit analysis that

we discuss them very generally again now.

Both lumped and distributed models involve the use of mathematical variables to represent physical quantities. Although the subject matter of this course is limited to lumped models, the basic nature of the variables is also applicable to distributed models. This should be expected since an electric current, for example, may equally be called upon to flow along the axis of a cylindrical conductor or along the surface of a sheet of conducting material. Basically there are only two types of physical quantities: those which flow along or through a constrained path, such as water flowing through a pipe, and those which imply a measurement which must be made between two points such as the distance between the points. In connecting a measuring instrument between two points it must be placed across everything that lies between them, and therefore such quantities are described as "across variables". In connecting an instrument to measure the flow along a path it is necessary to cut the path and insert the instrument so that the flow is constrained to pass through it, and therefore such quantities are described as "through variables".

The classification of all physical quantities being modelled as either "through" or "across" provides a powerful tool for the exploitation of analogous situations. Historically, these analogies become evident only after the mathematical model, in the form of a set of equilibrium equations, has been obtained. That is, it is possible for a particular mechanical system, fluid system, thermal system and electric circuit to have equilibrium equations which have identical mathematical form. Having obtained a solution for any one of them, the solutions for the others have effectively been obtained. The benefits resulting from these analogies can be realized more quickly if they can be recognized early in the modelling process. The concept of the through and across variables is the key to the formation of the analogous circuit diagram which effectively forms a

bridge between the physical model and the detailed mathematical model. This diagram has the appearance of an
electric circuit. The main advantage of the analogous
circuit diagram is that each different type of system is
modelled by a diagram having the same general appearance.
As a result, the process of obtaining a set of equilibrium
equation for any physical circuit is a straight-forward
technique which is very simple to apply, and has the effect
of providing a unified approach to the analysis of electrical, mechanical, fluid and thermal systems.

   In the succeeding chapters, the details of this approach
are developed. The through and across variables are identified for systems involving fluid flow, electric circuits,
mechanical translation, mechanical rotation, and heating
problems (Chapter III). After the process of obtaining the
analogous circuit has been discussed (Chapter IV) some very
basic physical situations are considered. Thereafter the
analytical techniques (Chapters V and VI) are developed
with examples drawn from any or all of these types of
systems so that by the end of the course a student should
have equal skill and confidence when faced with problems
of modelling and analysis of any of the systems included
in this study.

CHAPTER 2

SINGULARITY FUNCTIONS

In general, a real system is subject to a wide variety
of random input functions. The system operates on these
inputs to produce the output response. However, in the
design and analysis of systems it is usually impossible to
know in advance the exact input function that the system
will experience in service. For example, consider an automobile (Figure 2.1) as it approaches a bump in the road.
Let us assume that the automobile has a constant forward
velocity, v_x. After the front wheels reach the bump a
transverse velocity, v_y, is imparted to the body through
the springs and shock absorbers. This transverse velocity
is an input function to the car suspension. The magnitude
of such an input depends on the size and shape of the bump

and also on the forward speed of the car. Designers of
cars cannot know exactly what kind of roads will be encountered or how fast each driver will choose to drive.
For this reason it is convenient to use a set of idealized
input functions. Thus if two automobile suspensions are
designed we may determine which one gives the better ride
for a prescribed idealistic input function. In other words,
these idealistic input functions provide us with a means of
comparing the responses of systems. Furthermore, we may
formulate more realistic input functions from linear combinations of the idealistic input functions.

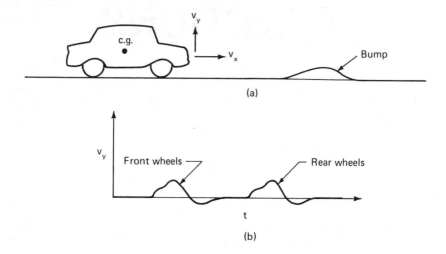

Figure 2.1 Transverse velocity, v_y, of car going over bump

In this chapter, we will introduce most of the standard
singularity functions which are used to express these
input functions in mathematical form.

Basically a singularity function is used to describe any
sudden change in the amplitude, slope, or any higher order
derivative of a function. The discontinuity at a sudden
change in amplitude is the most evident and is regarded
as a "step" as shown in the next section. A sudden change
in slope may not be as easily seen but nevertheless it is

very simply modelled by a ramp which will be shown to be
the integral of a step. Thus the singularity functions are
related to each other by either differentiation or integration.

2.1 THE UNIT STEP FUNCTION

The most common singularity function is the unit step
function. This type of function is often approximated in
real life when we switch on the lights or turn on a water
tap in a house. The step represents a sudden change in
the input from one level to another. We may get a better
understanding of a step change by referring to Figure 2.2.
Initially (Figure 2.2a) a butterfly valve blocks the exit
line from a water tank. With the valve closed there is no
flow, q, from the tank. If the valve is suddenly opened
water flows from the tank. Figure 2.2b shows the butterfly
valve in its fully opened position. The valve, however,
is a mechanical part and cannot move from the closed to
open position instantaneously. This time restriction on
sudden changes applies to all physical systems. The time
required to open the valve is reflected in the shape of
the actual flow, q(t), from the tank (Figure 2.2c). Note
also that in this example the flow decreases slightly
after the valve is fully opened. After sufficient time
the tank will be empty and the flow reduced to zero. Thus
the exact flow from the tank depends on the valve opening
time and the size of the tank.
The function we use to approximate sudden changes is the
unit step function, $u_s(t)$, shown in Figure 2.3. In this
function an abrupt change takes place in zero time. That
is the function goes from the zero level to the one level
instantaneously. We express this function in mathematical
form as:

$$u_s(t) = 1, \quad \text{for} \quad t > 0$$
$$= 0, \quad \text{for} \quad t < 0 \qquad (2.1)$$

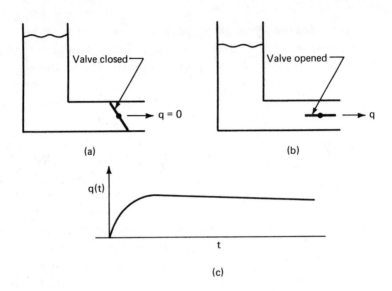

(a)　　　　　　　　　　　　　　(b)

(c)

Figure 2.2　Flow from a tank through a quick-opening valve

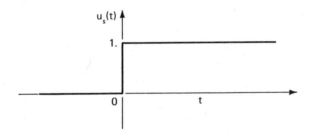

Figure 2.3　The unit step function

The argument of a particular step function is the quantity in parenthesis. In Equation (2.1) the argument is
the time, t. However, the function holds for other arguments. In other words the unit step is a function that
has a value of unity when the argument is greater than
zero and has a value of zero when the argument is less
than zero. The function is not defined when the argument
is exactly equal to zero.

As a consequence of the definition given above we may shift the unit step along the time axis by changing the argument. Figure 2.4 shows, for example, the advance and delay of the unit step function. Remember that the change always occurs when the argument is zero. Therefore, to advance the step function T time units we use the argument, t + T, with the result that:

$$
\begin{aligned}
u_S(t + T) &= 1, \quad \text{for} \quad t + T > 0 \quad \text{or} \quad t > -T \\
&= 0, \quad \text{for} \quad t + T < 0 \quad \text{or} \quad t < -T \qquad (2.2)
\end{aligned}
$$

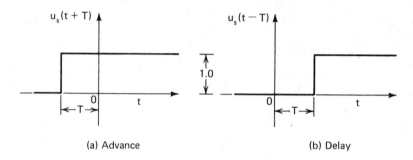

(a) Advance   (b) Delay

Figure 2.4 Advancing and delaying the unit step function

In a similar way we may delay the step function T time units by using the argument t - T. That is:

$$
\begin{aligned}
u_S(t - T) &= 1, \quad \text{for} \quad t - T > 0 \quad \text{or} \quad t > T \\
&= 0, \quad \text{for} \quad t - T < 0 \quad \text{or} \quad t < T \qquad (2.3)
\end{aligned}
$$

The unit step function has only two possible values, zero and unity. However, we may represent step changes of other amplitudes A in the form $Au_S(t)$. This means that:

$$
\begin{aligned}
Au_S(t) &= A, \quad \text{for} \quad t > 0 \\
&= 0, \quad \text{for} \quad t < 0 \qquad (2.4)
\end{aligned}
$$

It is also possible to combine step functions algebraically to attain composite functions of greater complexity. Two
types of problems are of particular interest. In one, we
decompose a function presented graphically into step function parts and then express the function in mathematical
form by summing the parts. In the other, we may need to
interpret mathematical formulation by plotting the function. We will consider one problem of each type in the
following two examples.

Example 2.1

We are given the rectangular square pulse shown in
Figure 2.5 and asked to determine, g(t), in mathematical
form.

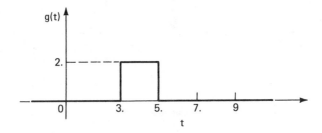

Figure 2.5   Rectangular square pulse (example 2.1)

Our first step must be to decompose the given pulse
into individual parts whose mathematical form we know.
From the given function we may observe that there is a
sudden change from 0 to 2 at t = 3 seconds. We may represent this sudden change by the step function,
$2u_s(t - 3)$, shown in Figure 2.6. This function remains
at the level of two for all times greater than 3 seconds.
The given function, on the other hand, is equal to zero
for all times greater than 5 seconds. We can obtain
this result by adding a step of -2 units at 5 seconds.
This sudden decrease may be written mathematically as
$-2u_s(t - 5)$. The given function is seen to be simply

the sum of the component parts shown in Figure 2.6.
Thus the function given in Equation (2.5) is:

$$g(t) = 2u_s(t - 3) - 2u_s(t - 5)$$

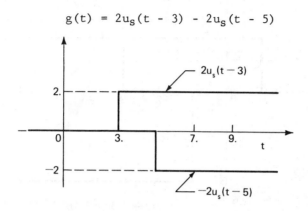

Figure 2.6 Decomposition of square pulse (example 2.1)

Example 2.2

Show the graphical representation of the function,
g(t), where

$$g(t) = 3u_s(t + 1) - 5u_s(t - 1) + 4u_s(t - 3)$$

In this type of problem we must plot each segment
separately and then add the ordinates algebraically
at each value of time. Figure 2.7 shows the graphical
representation of the three individual segments that
comprise the given function. Now to obtain g(t) we must
add the individual segments. To avoid confusion it is
a good policy to do the summation immediately before and
after each sudden change. Thus at times slightly less
than one second we are adding amplitudes of 3, 0, and 0.
At times slightly greater than one second we are adding
amplitudes of 3, -5, and 0. Thus the level changes from
+3 to -2 at a time of one second. This is shown in the
reconstruction in Figure 2.8. There is another sudden
change at 3 seconds. When time is slightly greater than

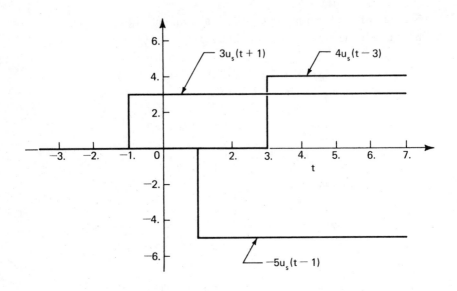

Figure 2.7 Segments of function (example 2.2)

3 seconds we must add amplitudes 3, -5, and 4. As a
result the level changes from -2 to +2 at a time of 3
seconds. Remember that we must always add the same
number of segments as there are terms in the given function. In this example there were three terms and this
produced three quantities to add. The complete function
is shown in Figure 2.8.

2.2 FUNCTIONS DERIVED BY INTEGRATION OF UNIT STEP

A single integration of the unit step function with
respect to time results in the unit ramp function, $u_r(t)$.
That is:

$$u_r(t) = \int_{-\infty}^{t} u_s(t')dt' \qquad (2.5)$$

The auxiliary variable, t', is introduced since the
upper limit of integration is to be the original independent variable, t.

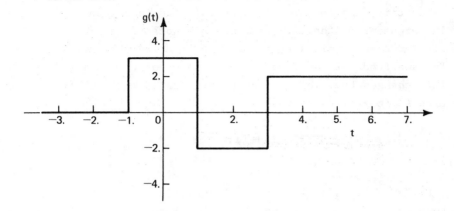

Figure 2.8 Reconstruction of function (example 2.2)

If we substitute the unit step function, Equation (2.1), into Equation (2.5) and perform the integration we may express the unit ramp function as:

$$u_r(t) = t, \quad \text{for} \quad t \geq 0$$
$$= 0, \quad \text{for} \quad t \leq 0 \tag{2.6}$$

Note that $u_r(t)$ is defined correctly for $t = 0$ by either condition.

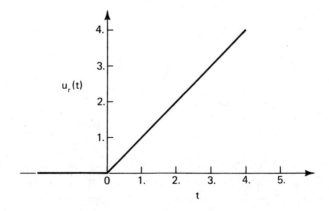

Figure 2.9 The unit ramp function

The unit ramp function is shown plotted in Figure 2.9. The argument of the unit ramp may be replaced in a manner similar to the unit step to advance or delay the function. The amplitude of the ramp may also be other than unity. This changes the slope of the ramp. For example the function $Au_r(t + T)$ is defined as:

$$Au_r(t + T) = A(t + T), \quad \text{for} \quad t + T \geq 0$$
$$= 0 \quad\quad\quad , \quad \text{for} \quad t + T \leq 0 \qquad (2.7)$$

Figure 2.10 is a plot of the function described in Equation (2.7) with a time advance of 1 second (T = 1) and an amplitude A = 1/2. In functional form we may write:

$$(1/2)u_r(t + 1) = (t + 1)/2 \ , \quad \text{for} \quad t \geq -1$$
$$= 0 \quad\quad\quad , \quad \text{for} \quad t \leq -1 \qquad (2.8)$$

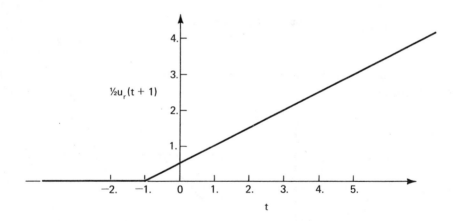

Figure 2.10 The ramp function advanced

Some composite functions may be separated into a sum of ramp functions with different arguments. Here again, we must use decomposition to identify the basic segments.

This may be done by noting that the addition of two straight
lines of the form $y_1 = m_1x + b_1$ and $y_2 = m_2x + b_2$ yields a
straight line of the form $y = (m_1 + m_2)x + (b_1 + b_2)$. The
gradient of the summed relation is the sum of the gradients.
Thus when adding or subtracting ramp functions we merely
add or subtract their amplitudes. The following example
will clarify the procedure.

Example 2.3

Figure 2.11a shows a function which begins like a
ramp delayed and then levels off. This is a useful
input function since switches and valves do not change
levels instantaneously (refer to Figure 2.2). The
problem is to describe this function in terms of the
known basic functions.

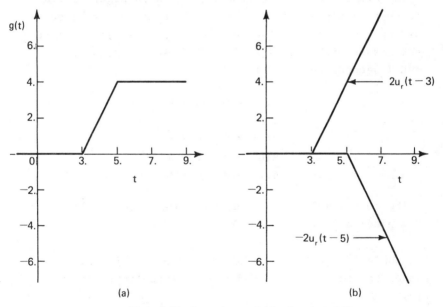

(a)   (b)

Figure 2.11 Ramp decomposition (example 2.3)

To decompose the given function the initial ramp
portion suggests a ramp function delayed. Figure 2.11b

shows this portion identified as $2u_r(t - 3)$. When $t = 5$ seconds this portion has a magnitude of 4 and this is the correct levelling off magnitude. However, the function $2u_r(t - 3)$ continues to rise. At $t = 7$ seconds it has a magnitude of 8. Thus the function $2u_r(t - 3)$ only describes the given function for $t \leq 5$ seconds. We must therefore, subtract a second function beginning at $t = 5$ seconds. This function must have the same slope as the initial ramp portion and be opposite in sign. The composite function will then have zero slope. This second portion is therefore $-2u_r(t - 5)$. The given composite function may be described as:

$$g(t) = 2u_r(t - 3) - 2u_r(t - 5)$$

If this is really the function given in Figure 2.11a, it must be correct for all times. We may check by evaluating the function at a few specific times and comparing the result with Figure 2.11a. Thus at $t = 3, 4, 5, 6,$ and 7 seconds the function is:

$$g(3) = 0 - 0 = 0$$
$$g(4) = 2 - 0 = 2$$
$$g(5) = 4 - 0 = 4$$
$$g(6) = 6 - 2 = 4$$
$$g(7) = 8 - 4 = 4$$

The functional form yields the same values as Figure 2.11a for all times.

A double integration of the unit step function with respect to time results in the unit parabolic function $u_p(t)$:

$$u_p(t) = \int_{-\infty}^{t} \int_{-\infty}^{t'} u_s(t'')dt''dt' \qquad (2.9)$$

The unit parabolic function may also be defined in terms of the unit ramp function as:

$$u_p(t) = \int_{-\infty}^{t} u_r(t')dt' \qquad (2.10)$$

If we substitute the functional forms for the unit ramp (given in Equation (2.6)) into Equation (2.10), the unit parabolic function becomes:

$$\begin{aligned} u_p(t) &= t^2/2, \quad \text{for} \quad t \geq 0 \\ &= 0 \quad , \quad \text{for} \quad t \leq 0 \end{aligned} \qquad (2.11)$$

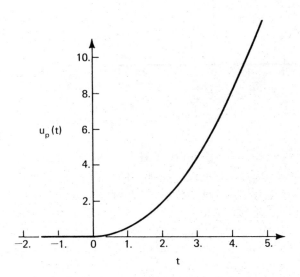

Figure 2.12 Unit parabolic function

The unit parabolic function is shown in Figure 2.12. As in the step and ramp functions we may modify the amplitude and argument of the parabolic function. The function $Au_p(t - T)$ is defined as:

$$\begin{aligned} Au_p(t - T) &= A(t - T)^2/2, \quad \text{for} \quad t \geq T \\ &= 0 \quad\quad\quad\quad , \quad \text{for} \quad t \leq T \end{aligned} \qquad (2.12)$$

Figure 2.13 shows the parabolic function $4u_p(t - 2)$.

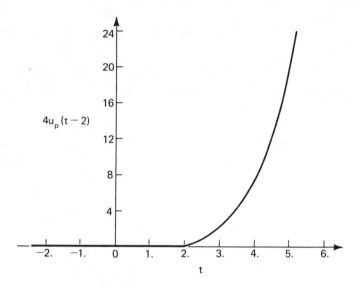

Figure 2.13 Parabolic function delayed

The addition and subtraction of parabolic functions requires somewhat more care than with ramp or step functions. The operation may proceed directly from the definition of the parabolic functions. Another approach, however, is to differentiate the composite parabolic function. The resulting ramp functions may be treated as previously indicated and then converted back to the parabolic function by integration. The following example demonstrates the procedures.

Example 2.4

Plot the function described by

$$g(t) = u_p(t) - 2u_p(t - 1) + u_p(t - 2)$$

a) We merely plot each component of the given function separately and then add. Figure 2.14a shows the three segments of the given function plotted separately. To compute the function rather than depend on graphical addition it is convenient to rewrite the given function as:

$$g(t) = u_p(t) = t^2/2 + 0 + 0, \text{ for } 0 \le t \le 1$$

$$g(t) = u_p(t) - 2u_p(t - 1) + 0 = \frac{t^2}{2} - (t - 1),$$
$$\text{for } 1 \le t \le 2$$

$$g(t) = u_p(t) - 2u_p(t - 1) + u_p(t - 2)$$
$$= \frac{t^2}{2} - (t - 1)^2 + \frac{(t - 2)^2}{2}, \text{ for } t \ge 2$$

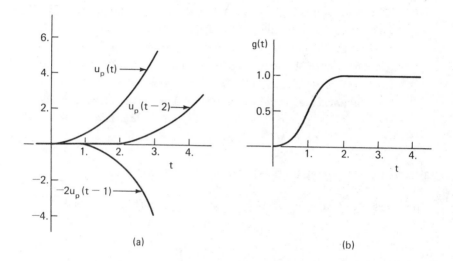

(a)   (b)

Figure 2.14 Development of function for example 2.4

From this formulation we may compute the value of the function at any time. Thus:

$$g(1.5) = 1.125 - 0.250 = 0.875$$

$$g(3.5) = 6.125 - 6.250 + 1.125 = 1.000$$

Figure 2.14b shows the required function.
b) The other approach is to differentiate the given composite function. Thus we obtain:

$$g'(t) = u_r(t) - 2u_r(t - 1) + u_r(t - 2)$$

This function represents a triangular pulse. We
may describe g'(t) functionally and integrate the
results to obtain the same functions of g(t) that
were derived in part a). However, we may also
perform a graphical integration and get the desired
function of g(t) directly. In the graphical integration procedure we find the area under the function from time zero to the time under consideration.
For example if we wanted to know g(1) we would find
the area under the g'(t) function from time, t = 0
to t = 1. The result would be that g(1) = 0.5. If
on the other hand we wanted g(3) we would need the
area from t = 0 to t = 3. In this case it is only
the area under the triangular function from t = 0
to t = 2. There is no additional area from t = 2
to t = 3, thus g(2) = g(3) = 1.0.

2.3 FUNCTIONS DERIVED BY DIFFERENTIATION OF UNIT STEP

The time derivative of the unit step function is the
unit impulse function, $u_i(t)$ or in mathematical terms:

$$u_i(t) \equiv \frac{d}{dt} [u_s(t)] \tag{2.13}$$

We may evaluate the unit impulse function by substituting
the values of the unit step (Equation (2.1)) into Equation
(2.13). The result is:

$$
\begin{aligned}
u_i(t) &= 0, \text{ for } t > 0 \\
&= 0, \text{ for } t < 0 \\
&= \infty, \text{ for } t = 0
\end{aligned}
\tag{2.14}
$$

Figure 2.15 shows a graphical representation of the unit
impulse function. A useful property of the impulse function can be obtained by integrating both sides of Equation
(2.13) with respect to time so that

$$\int_{-\infty}^{+\infty} u_i(t)\,dt = u_s(t) \Big|_{-\infty}^{+\infty} = 1 \tag{2.15}$$

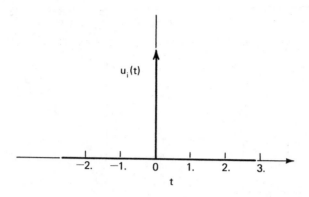

u_i(t)

Figure 2.15 Graphical representation of unit impulse function

Thus the area bounded by the unit impulse function and the time axis is always equal to unity. An alternate approach to the impulse function is through a limiting process. Consider the function $g(t)$ shown in Figure 2.16a. We may express this function in terms of ramp functions as:

$$g(t) = \frac{1}{T} [u_r(t) - u_r(t - T)] \qquad (2.16)$$

Now note that as T becomes smaller $g(t)$ tends toward the unit step function. That is:

$$u_s(t) = \lim_{T \to 0} \frac{1}{T} [u_r(t) - u_r(t - T)] \qquad (2.17)$$

The advantage of this formulation is that we may perform the differentiation of Equation (2.17) which yields the unit impulse function as:

$$u_i(t) = \lim_{T \to 0} \frac{1}{T} [u_s(t) - u_s(t - T)] \qquad (2.18)$$

Figure 2.16b shows the function described in Equation (2.18). As time approaches zero the amplitude approaches infinity. However, the area under the impulse still remains unity. If we wish to designate functions with

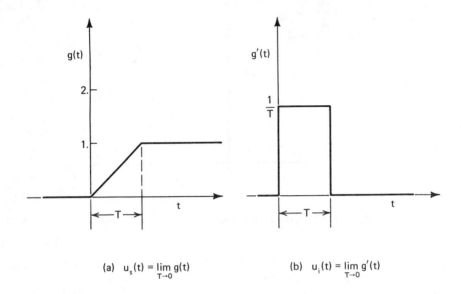

(a) $u_s(t) = \lim_{T \to 0} g(t)$ (b) $u_i(t) = \lim_{T \to 0} g'(t)$

Figure 2.16 The impulse function as a limit

other than unit area we may do so by multiplying the
impulse by the appropriate constant (i.e. $Au_i(t)$). Thus
the function

$$g(t) = 5u_i(t - 1) + 2u_i(t - 3) + 3u_i(t - 4) \qquad (2.19)$$

which represents a particular series of impulse functions
may be represented graphically as shown in Figure 2.17.
The "strength" of each impulse is indicated by the number
at the side and refers to the area under the impulse.

   The second time derivative of the unit step function is
the unit doublet function $u_d(t)$

$$u_d(t) \equiv \frac{d^2}{dt^2} [u_s(t)] = \frac{d}{dt} [u_i(t)] \qquad (2.20)$$

   Let us use the limiting approach to determine values for
the doublet function. As a first step we construct a function which approaches the impulse function in the limit.

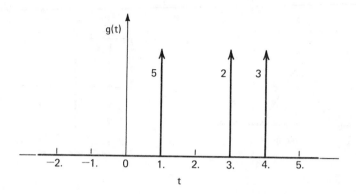

Figure 2.17 Graphical representation of a series of impulse functions

The desired function must also enclose a unit area. In the construction we use segments that are readily differentiated. Figure 2.18a shows the function $g_1(t)$ which may be expressed as:

$$g_1(t) = \frac{1}{T^2} [u_r(t + T) - 2u_r(t) + u_r(t - T)] \qquad (2.21)$$

To verify that Equation (2.21) is the mathematical expression for Figure 2.18a the student should apply Equation (2.6). Now note that the unit impulse function is:

$$u_i(t) = \lim_{T \to 0} g_1(t) \qquad (2.22)$$

From Equations (2.20), (2.21) and (2.22) the unit doublet function may be written as:

$$u_d(t) = \lim_{T \to 0} g_1'(t)$$

$$= \lim_{T \to 0} \frac{1}{T^2} [u_s(t + T) - 2u_s(t) + u_s(t - T)] \qquad (2.23)$$

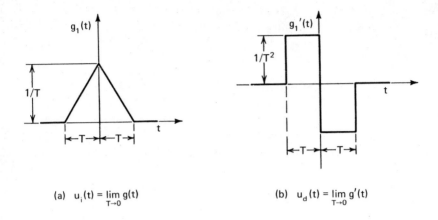

(a) $u_i(t) = \lim_{T \to 0} g(t)$ (b) $u_d(t) = \lim_{T \to 0} g'(t)$

Figure 2.18 The unit doublet function as a limit

Figure 2.18b shows the limiting function that represents the unit doublet function in Equation (2.23). As T approaches zero the unit doublet may be evaluated as:

$$
\begin{aligned}
u_d(t) &= 0, &&\text{for } t < 0^- \\
&= 0, &&\text{for } t > 0^+ \\
&= +\infty, &&\text{for } t = 0^- \\
&= -\infty, &&\text{for } t = 0^+ \qquad (2.24)
\end{aligned}
$$

where the designations 0^- and 0^+ refer to an instant before and after $t = 0$. The graphical representation of the unit doublet function is shown in Figure 2.19.

2.4 EXPONENTIAL FUNCTIONS

The exponential functions that we are about to consider are not singularity functions. However, they are often combined with singularity functions to describe system inputs. It is therefore convenient to include them at this point.

The most general form of the exponential function is:

$$g(t) = e^{st} \qquad (2.25)$$

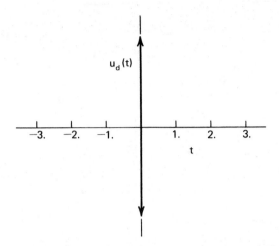

Figure 2.19 Graphical representation of unit doublet function

where s may be a complex number $\sigma + j\omega$. We will consider
only two special cases of the exponential function:
1) when s is purely real ($\omega = 0$) and
2) when s is purely imaginary ($\sigma = 0$).
When s is only a real number Equation (2.25) becomes:

$$g(t) = e^{\sigma t} \qquad\qquad (2.26)$$

Figure 2.20 is a plot of Equation (2.26) in which σ has
the values +.3 and -.3. When the exponent, σ, is negative
the function is known as the exponential decay. A positive
exponent results in the exponential rise function. For the
special case where $\sigma = 0$, the function has the value of
unity for all times.
   When s is a purely imaginary number Equation (2.25) is
written as:

$$g(t) = e^{j\omega t} \qquad\qquad (2.27)$$

Figure 2.21 shows the representation of the exponential
function in Equation (2.27) as a rotating unit vector.
The angular frequency of the rotation depends on the

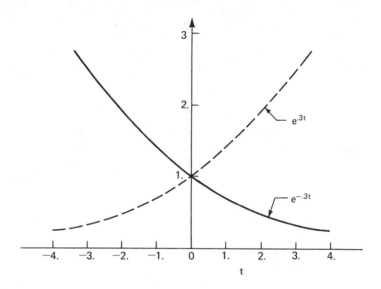

Figure 2.20 Exponential decay and rise functions

magnitude of ω. The vector rotates counterclockwise as
time increases. According to Euler's formula the imaginary exponential may be expressed as:

$$e^{j\omega t} = \cos\omega t + j \sin\omega t \qquad (2.28)$$

If we use the symbols Re[] and Im[] to indicate the real
and imaginary parts of the bracketed quantity then Equation
(2.28) may be used to define the circular functions, $\cos\omega t$
and $\sin\omega t$.

$$Re(e^{j\omega t}) = \cos\omega t$$
$$Im(e^{j\omega t}) = \sin\omega t \qquad (2.29)$$

Equation (2.29) may be more readily visualized by comparing the unit vector components with the inserted triangle
on Figure 2.21. If we plot the functions given in Equation
(2.29) against a linear time scale the familiar cosine and
sine shapes appear. Figure 2.22 shows this plot.

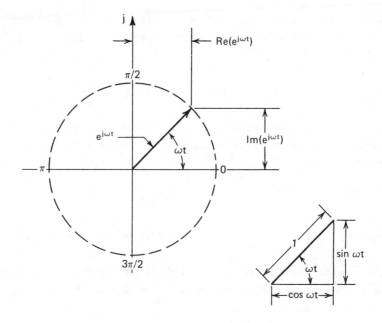

Figure 2.21 Representation of exponential as rotating vector

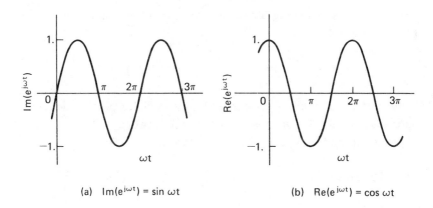

(a) $Im(e^{j\omega t}) = \sin \omega t$ (b) $Re(e^{j\omega t}) = \cos \omega t$

Figure 2.22 Real and imaginary parts of exponential function

Example 2.5

In the solution of problems we often obtain functions that are the sums or differences of sine and cosine terms

of the same frequency. We may always express such functions in terms of sine or cosine functions alone. Suppose
we have

$$g(t) = A \sin\omega t + B \cos\omega t$$

It is not easy to interpret this function directly and
therefore it usually is essential to convert it to the
form

$$g(t) = D \sin(\omega t + \theta_1)$$

where θ_1 is a constant angle and D is a constant amplitude.

The usual way to accomplish this is to apply the double
angle formula of trigonometry so that $D \sin(\omega t + \theta_1)$ is
expanded to:

$$(D \cos\theta_1)\sin\omega t + (D \sin\theta_1)\cos\omega t = A \sin\omega t + B \cos\omega t$$

By equating the coefficients of the $\sin\omega t$ and $\cos\omega t$
terms we find that:

$$D = \sqrt{A^2 + B^2} \quad , \quad \theta_1 = \tan^{-1}(B/A)$$

We may obtain this same result by working with complex
numbers. Thus if we desire $g(t) = D \sin(\omega t + \theta_1)$ we may
start by putting the given equation in the form:

$$\text{Im}\left[D \exp[j(\omega t + \theta_1)]\right] = \text{Im}\left[A \; e^{j\omega t}\right] + \text{Re}\left[B \; e^{j\omega t}\right]$$

However, we can only add directly if all the terms are
either Im or Re. This can be changed to that form by
rotating the last terms on the right so that

$$\text{Im}\left[D \exp[j(\omega t + \theta_1)]\right] = \text{Im}\left[A \; e^{j\omega t}\right] + \text{Im}\left[jB \; e^{j\omega t}\right]$$

Now since all the terms are Im we may drop this designation and

$$D \, e^{j\theta_1} e^{j\omega t} = A \, e^{j\omega t} + jB \, e^{j\omega t}$$

In polar form this reduces to:

$$D \, e^{j\theta_1} = \sqrt{A^2 + B^2} \ \exp(j \, \tan^{-1} B/A)$$

from which

$$D = \sqrt{A^2 + B^2} \quad \text{and} \quad \theta_1 = \tan^{-1} B/A$$

The procedure to place the expression in the form $g(t) = D \cos(\omega t - \theta_2)$ is similar and is left as an exercise for the student.

2.5 COMPOSITE FUNCTIONS

The basic singularity functions may be combined to describe a wide variety of time functions. These composite functions are usually formed by adding or multiplying basic functions. In the case of multiplication the unit step and the unit step shifted are most often used. We must be able to find both the functional representation from the graphical time plot and the inverse process. Examples 2.6 and 2.7 will demonstrate some composite functions and their formulation.

Example 2.6

We are given the function shown in Figure 2.23 and asked to find an expression which represents it.

The procedure to follow in this type of problem is to start at t = 0 and see what has to be done at this time. From the function we note that at t = 0 there is a change from 0 to 2. This means immediately that one of the com-

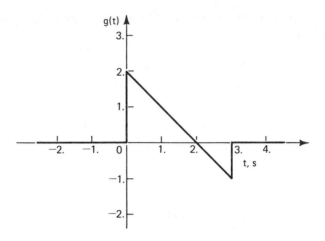

Figure 2.23 Function for example 2.6

ponents of the required function is $2u_s(t)$ because in no
other way can we duplicate the sudden change at t = 0.
Now the function magnitude decreases from t = 0^+ to
t = 3. The decrease, however, begins at t = 0^+ and for
this reason we must begin subtracting from the step function at this time. The slope of the given function between
t = 0 and t = 3 is -1 and this suggests we subtract a unit
ramp, $u_r(t)$. If we combine these two functions we obtain
the function $2u_s(t) - u_r(t)$. To see what we have so far,
refer to Figure 2.24a. Comparison of Figure 2.24a with
the desired function shows that they are the same until
t = 3. However, at this time the given function again
changes suddenly from -1 to 0. To accomplish this change
we must add a unit step shifted by 3 seconds. Remember
that the function formed up to 3 seconds remains in effect
for all time. Thus by adding the step delayed by 3 seconds
we have accumulated a function in the form $2u_s(t) - u_r(t)$
+ $u_s(t - 3)$. This function is shown in Figure 2.24b. If
we again compare the function with Figure 2.23 we see that
our combined function decreases for times greater than
three seconds, whereas the desired function remains at

zero. This means that we must add another function beginning at t = 3 to counteract the decrease of the unit ramp.
The required component is a unit ramp which starts increasing at t = 3. Thus the complete function which describes the function in Figure 2.23 is:

$$g(t) = 2u_s(t) - u_r(t) + u_s(t - 3) + u_r(t - 3)$$

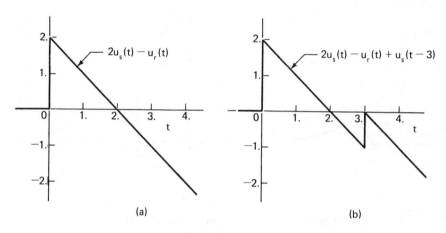

Figure 2.24 Stages in building up function for example 2.6

We limit our consideration of the multiplication of two
time functions to cases where one of the functions is a
combination of unit step functions. Thus, in effect, we
will be multiplying a given function, g(t), by 0 or 1 in
specific time intervals. The resulting functional product
will then represent the given function g(t) with sections
deleted. Suppose we wish to express an exponential decay
that holds only when t > 0. To do this we would use the
product form $e^{\sigma t} u_s(t)$. Figure 2.25a shows the result.
The step function effectively blocks out the function for
t < 0. To shift this exact function in time we need only
replace all values of t by (t ± T). Thus to delay $e^{\sigma t} u_s(t)$
by T we would obtain $e^{\sigma(t-T)} u_s(t-T)$ as the shifted function
(Figure 2.25b). Another way of interpreting this function

is as $(e^{-\sigma T})e^{\sigma t}u_s(t-T)$. Instead of shifting the function
of Figure 2.25a we could have blocked out an exponential
with amplitude, $e^{-\sigma T}$, for all times less than T.

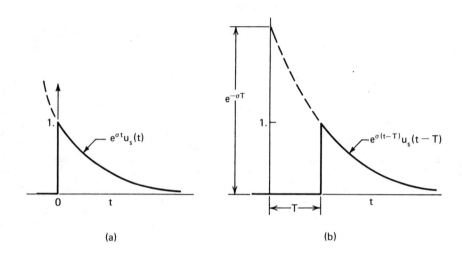

Figure 2.25   Effect of multiplication by unit step function

Example 2.7

The function shown in Figure 2.26a is one half cycle
of a sinusoid.   Find the functional representation.

The required function h(t) will be the product of two
time functions.   One of the functions must obviously be
the sine function.   Thus we may write:

$$h(t) = g(t) \sin t$$

We must determine the other time function g(t), to
produce the given function.   The characteristics of g(t)
must therefore be

$$g(t) = 1, \quad \text{for} \quad 0 < t < \pi$$
$$ = 0, \quad \text{for} \quad \text{all other values of t.}$$

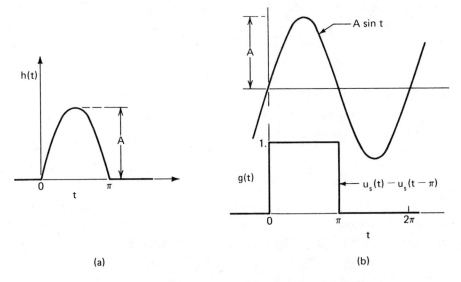

Figure 2.26 Multiplication of functions (example 2.7)

Figure 2.26b is a graphical representation of $g(t)$. To obtain this function we must add a unit step and a negative unit step delayed by π. The result is $g(t) = u_s(t) - u_s(t - \pi)$ and the required function is:

$$h(t) = [u_s(t) - u_s(t - \pi)] \sin t$$

2.6 MATHEMATICAL OPERATIONS ON FUNCTIONS

In the previous sections we have considered a broad class of input functions, $g_i(t)$. We now turn our attention on the production of output functions, $g_o(t)$, by mathematical operations on the input functions. Figure 2.27 is a block diagram representation of this process in which an input function is operated on to produce an output function. We deal here with four common operations that are performed by physical components and systems. For the time being, however, we will treat the operations as mathematical exercises

without regard to their physical implementations. The
operations are:

1) Time Integration; 2) Time Differentation; 3) Proportional plus Integral; and 4) Proportional plus Derivative.

Time Integration

When the mathematical operation is integration the output
function is related to the input function by:

$$g_o(t) = \int g_i(t)dt \qquad (2.30a)$$

Recall that with an indefinite integral there is a constant of integration that must be evaluated. However,
integrations performed by real physical components are
always definite integrations that reflect the entire
history of the components. Thus in this context the

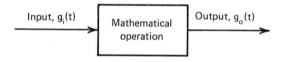

Figure 2.27 Block diagram representation of input-output relation

operation of integration should be expressed as:

$$g_o(t) = \int_{-\infty}^{t} g_i(t')dt' \qquad (2.30b)$$

where t' is a dummy variable. Equation (2.30b) may be conveniently separated into two component integrals:

$$g_o(t) = \int_{-\infty}^{t_b} g_i(t')dt' + \int_{t_b}^{t} g_i(t')dt' \qquad (2.30c)$$

or

$$g_o(t) = g_o(t_b) + \int_{t_b}^{t} g_i(t')dt' \qquad (2.30d)$$

where t_b is the time at the beginning of the operation and
$g_o(t_b)$ is the value of the output function at $t = t_b$. Equation (2.30d) is therefore the basic form of integration
which is appropriate for physical systems.

An important feature of the integration operation is
that the output function always depends on both the integration and the previous value of the output. Thus integrating devices have memory. To demonstrate this let us
return to the operation indicated in Equation (2.30d).
Suppose that the input function in a particular interval
has a zero value, the output function then retains its
last value and thus reflects the effects of all previous
inputs.

To visualize the mathematical operation it is often
advantageous to employ graphical methods. Thus, in the
case of integration the output function at any time will
equal the area under the input function up to the time
under consideration.

The following example will make the integration process
clear.

Example 2.8

The input function shown in Figure 2.28a is fed into
an integrator. Determine the resulting output function
if $g_o(0) = 0$.

There are several ways of finding the output function.
We may use either a mathematical or graphical approach.
In this example we will demonstrate each method.

a) Mathematical Integration

The most direct way to perform mathematical integration
is to break the input function into an appropriate number
of intervals in which the functional form remains the
same. That is:

$$1\underline{st} \text{ interval} \rightarrow g_i(t) = 4 \quad 2 < t < 4$$
$$2\underline{nd} \text{ interval} \rightarrow g_i(t) = 0 \quad 4 < t < 8$$

3rd interval → $g_i(t)$ = 1   8 < t < 12
4th interval → $g_i(t)$ = 0   12 < t < 14
5th interval → $g_i(t)$ = -3   14 < t < 16

Now we proceed by applying Equation (2.30d) in each
interval in turn beginning with the first interval.
Thus:

$$g_o(t) = \int_2^t 4dt + 0 = 4t - 8, \quad 2 \le t \le 4$$

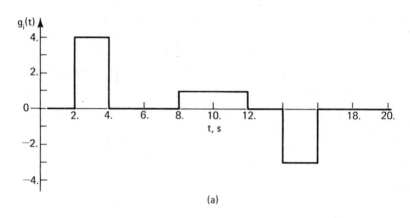

(a)

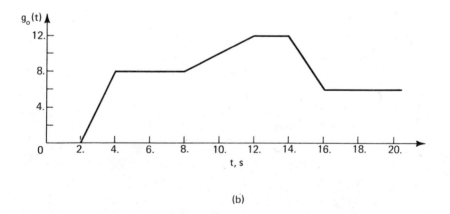

(b)

Figure 2.28   Integration of time functions (example 2.8)

We may use the equation to plot the function in the first interval. Figure 2.28b shows the output. At the end of the first interval $t = 4$ so that $g_0(4) = 8$. In the second interval the input function is zero. Thus the output throughout this interval remains at $g_0(4) = g_0(8) = 8$. In the third interval the output has the form:

$$g_0(t) = \int_8^t (1)dt + g_0(8) = t - 8 + 8$$
$$= t, \quad 8 \le t \le 12$$

We find values of the output in the third interval from this relation. At the end of the third interval the output $g_0(12) = 12$. Since there is no input in the fourth interval the output remains at the last previous level. Thus $g_0(14) = g_0(12) = 12$. Finally, the last interval has an output of:

$$g_0(t) = \int_{14}^t (-3)dt + g_0(14) = 42 - 3t + 12$$
$$= 54 - 3t, \quad 14 \le t \le 16$$

This relation holds for the output in the fifth interval. The output then stays at $g_0(16) = 6$ until another input is applied.

Another way of solving this problem is to express the entire input function in mathematical form. Thus:

$$g_i(t) = 4u_s(t - 2) - 4u_s(t - 4) + u_s(t - 8)$$
$$- u_s(t - 12) - 3u_s(t - 14) + 3u_s(t - 16)$$

If $g_i(t)$ is placed into Equation (2.30d) the output $g_0(t)$ is:

$$g_0(t) = 4u_r(t - 2) - 4u_r(t - 4) + u_r(t - 8)$$
$$- u_r(t - 12) - 3u_r(t - 14) + 3u_r(t - 16) + g(0)$$

A plot of the above function also yields the output shown in Figure 2.28b.

b) Graphical Integration

In this approach the output $g_o(t)$ is equal to the area under the $g_i(t)$ curve from 0 to t. For example, to find $g_o(4)$ we need the area under the input function from t = 0 to t = 4. This area is 4(4 - 2) = 8. Thus $g_o(4) = 8$.

To find $g_o(16)$ we must sum all the area under $g_i(t)$ from t = 0 to t = 16. Negative areas are subtracted from the total. Thus:

$$g_o(16) = 4(4 - 2) + 1(12 - 8) - 3(16 - 14) = 6$$

In those regions where the input has no value (i.e., for t between 4 and 8, or t between 12 and 14) the output function remains constant at its last previous value.

Time Differentiation

The relation between the output function and the input function when the mathematical operation is differentiation is:

$$g_o(t) = \frac{dg_i(t)}{dt} \qquad (2.31)$$

Differentiation emphasizes a change in the input function. The output does not depend on any previous inputs. Whenever the input function is not changing in value the output is zero. From a graphical viewpoint the output function is the slope of the input function at each instant of time.

Example 2.9

Figure 2.29a shows the input function supplied to a differentiator. Find and plot the corresponding output function.

a) Mathematical Differentiation

We will use two mathematical methods to find the output function. In the first method we express the given

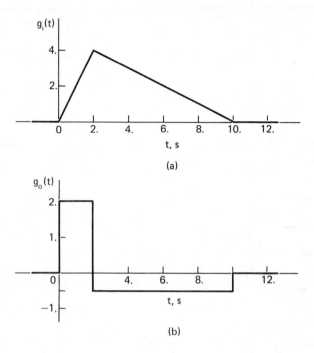

Figure 2.29 Differentation of time functions (example 2.9)

function in terms of time in specific intervals. For the given input function we may write:

$$g_i(t) = 2t \quad , \quad 0 < t < 2$$

$$g_i(t) = 5 - t/2, \quad 2 < t < 10$$

If these inputs are substituted into Equation (2.31) the output function is:

$$g_o(t) = 2 \quad , \quad 0 < t < 2$$

$$g_o(t) = -1/2, \quad 2 < t < 10$$

This result is shown in Figure 2.29b.

In the second method we write the input function in terms of time functions so that for this example:

$$g_i(t) = 2u_r(t) - (5/2)u_r(t - 2) + (1/2)u_r(t - 10)$$

Now the application of Equation (2.31) leads immediately to:

$$g_o(t) = 2u_s(t) - (5/2)u_s(t - 2) + (1/2)u_s(t - 10)$$

A plot of this function yields the same result shown in Figure 2.29b.

b) Graphical Differentiation

The output function is the slope of the input function at each value of time. The slope from $t = 0$ to $t < 2$ is constant and equal to $4/2 = 2$. Thus $g_o(1) = 2$. At $t = 2$ the slope changes discontinuously from positive to negative. The value of the output is constant between $t > 2$ and $t < 10$ because the slope is constant in this region. At $t = 6$, for example, $g(6) = 4/(2 - 10) = -1/2$. Any horizontal segment of the input function results in a zero output function since the horizontal line has zero slope. Thus for times greater than 10 seconds $g_o(t) = 0$.

Proportional Plus Integral

In this case the output function is the sum of two operations on the input function. The output due to the proportional operation is designated as $g_{op}(t)$ and the one due to the integration as $g_{oi}(t)$. Thus the operation performed can be represented mathematically by:

$$g_{op}(t) = A_1 g_i(t) \tag{2.32a}$$

$$g_{oi}(t) = A_2 \int_{t_b}^{t} g_i(t')dt' + g_{oi}(t_b) \tag{2.32b}$$

$$g_o(t) = g_{op}(t) + g_{oi}(t) \tag{2.32c}$$

where A_1 and A_2 are constants. The output from the proportional plus integral operation differs from the integration operation. The output of the proportional plus integral operation retains only the portion due to integration. Thus, when the input function goes to zero in an interval the output remembers only the results of previous integrations of the input function.

Example 2.10

The input function of Figure 2.30a is subjected to proportional plus integral action according to the formula

$$g_o(t) = 2g_i(t) + \int g_i(t)dt$$

If the initial condition on the integrator is zero plot the output function. The input function may be separated into two intervals as:

$$g_i(t) = (2t/3) + 1, \quad 0 < t < 3$$

$$g_i(t) = 0 \quad , \quad t > 3$$

The proportional portion of the output may be obtained directly as:

$$g_{op}(t) = (4t/3) + 2, \quad 0 < t < 3$$

$$g_{op}(t) = 0 \quad , \quad t > 3$$

This portion of the output is shown in Figure 2.30c. The integral portion of the output is:

$$g_{oi}(t) = \int_0^t ((2/3)t' + 1)dt' = (1/3)t^2 + t, \quad 0 < t < 3$$

$$g_{oi}(t) = 6 , \quad t > 3$$

The integral operation on the input is shown in Figure 2.30b. Now we may add to obtain the complete output for both operations as

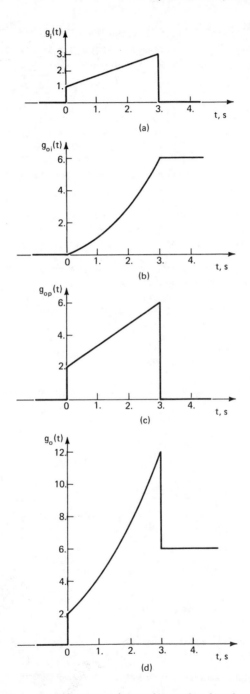

Figure 2.30 Proportional plus integral operation on time functions (example 2.10)

$$g_o(t) = (1/3)t^2 + (7/3)t + 2, \quad 0 < t < 3$$

$$g_o(t) = 6, \quad t > 3$$

Figure 2.30d shows the required output for proportional plus integral operation.

The same operations may be dealt with strictly in terms of time functions. Thus we begin with

$$g_i(t) = u_s(t) + (2/3)u_r(t) - 3u_s(t - 3) - (2/3)u_r(t - 3)$$

The output due to proportional operation is:

$$g_{op}(t) = 2u_s(t) + (4/3)u_r(t) - 6u_s(t - 3)$$
$$- (4/3)u_r(t - 3)$$

and that due to integral operation is:

$$g_{oi}(t) = u_r(t) + (2/3)u_p(t) - 3u_r(t - 3) - (2/3)u_p(t - 3)$$

The sum of these functions gives the same answer as shown in Figure 2.30d; however, the plot is more difficult to obtain. When the time functions are used for this problem, 8 separate segments must be considered.

The problem can also be solved using a strictly graphical approach.

Proportional Plus Derivative

This operation yields an output that depends on the sum of a proportional part, $g_{op}(t)$, and a derivative part, $g_{od}(t)$. The complete output may be expressed as:

$$g_o(t) = A_1 g_i(t) + A_2 \frac{dg_i(t)}{dt} \tag{2.33}$$

The operation produces an output that depends on the rate of change of the input to a lesser extent than in the differentiation operation alone. Thus the output is somewhat less sensitive to input changes.

Example 2.11

The input function given in Figure 2.31a, is applied at the input of a proportional plus derivative controller. Determine the output function if the mathematical operation is:

$$g_o(t) = 1.5g_i(t) + \frac{dg_i(t)}{dt}$$

1st interval → $g_i(t) = -2t$	,	$0 < t < 1$
2nd interval → $g_i(t) = -2$	,	$1 < t < 3$
3rd interval → $g_i(t) = 4t - 14$	,	$3 < t < 4$
4th interval → $g_i(t) = (-2/3)t + 14/3$,		$4 < t < 7$

Thus the output due to the derivative portion is:

$$g_{od}(t) = -2 \quad , \quad 0 < t < 1$$
$$g_{od}(t) = 0 \quad , \quad 1 < t < 3$$
$$g_{od}(t) = 4 \quad , \quad 3 < t < 4$$
$$g_{od}(t) = -2/3, \quad 4 < t < 7$$

This output portion is shown in Figure 2.31b. The proportional portion is merely the original input multiplied by 1.5 and is:

$$g_{op}(t) = -3t \quad , \quad 0 < t < 1$$
$$g_{op}(t) = -3 \quad , \quad 1 < t < 3$$
$$g_{op}(t) = 6t - 21, \quad 3 < t < 4$$
$$g_{op}(t) = -t + 7 \quad , \quad 4 < t < 7$$

Figure 2.31c shows the proportional part of the output. The complete output is the sum of the proportional and derivative parts.

$$g_o(t) = -3t - 2 \quad , \quad 0 < t < 1$$
$$g_o(t) = -3 \quad , \quad 1 < t < 3$$

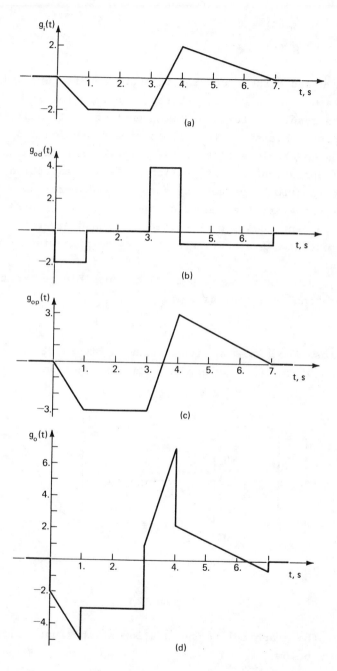

Figure 2.31 Proportional plus derivative operation on time functions (example 2.11)

$$g_o(t) = 6t - 17 \quad , \quad 3 < t < 4$$
$$g_o(t) = -t + 19/3, \quad 4 < t < 7$$

The required output function is shown in Figure 2.31d.
In this case also we may use the time functions. However, the result in terms of time functions will contain
10 separate functions. Thus it is time-consuming and
laborious to obtain a plot of the output. For the mathematical operations of integration and differentiation
the time function approach may sometimes be more convenient. However, the summation operations of proportional plus integral and proportional plus derivative
will almost always be easier by using the piecewise
approach.

Here, again, in this example, it is possible to treat
the operation by graphical means alone.

Problems

2.1 Express the time functions shown in Figure P2.1 in
terms of the appropriate singularity functions.

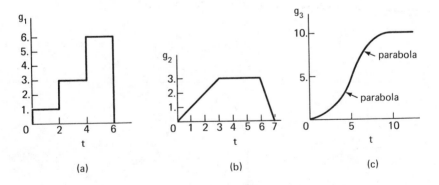

Figure P 2.1

2.2 Show the graphical representation of the functions
given below:

   (a) $g_1(t) = u_s(t + 1) - 2u_s(t) - 3u_s(t - 2)$

(b) $g_2(t) = -2u_r(t) - u_r(t - 2) + 6u_r(t - 4)$

(c) $g_3(t) = 2u_p(t - 2) - 2u_p(t - 5) - u_p(t - 8)$
$\quad\quad + u_p(t - 14)$

2.3 Sketch the graphs of the following expressions:

(a) $g_1(t) = u_s(t) + 2u_s(t - 2) + 5u_s(t - 3)$
$\quad\quad + u_r(t - 3) - u_r(t - 5)$

(b) $g_2(t) = u_r(t) - u_r(t - 1) + u_s(t - 2)$
$\quad\quad + (1/2)u_r(t - 2) + (1/2)u_r(t - 6)$

(c) $g_3(t) = 2u_p(t) - 36u_s(t - 4) - 2u_p(t - 6)$
$\quad\quad - 12u_r(t - 6)$

2.4 (i) Express the time functions shown in Figure
P2.4 in terms of the appropriate singularity functions.
(ii) Sketch the time integral of each function.

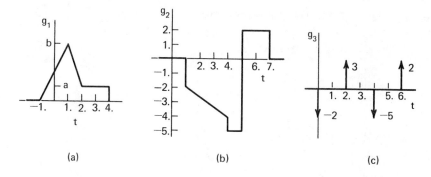

(a)   (b)   (c)

Figure P 2.4

2.5 Sketch the following function in the interval
$0 \le t \le 20$ s.

$g(t) = 5u_s(t) + u_r(t) + 0.2u_p(t) - 0.2u_p(t - 10)$

2.6 A four position valve is suddenly moved from position
    1 to position 2 where the water flow is 30 cm³/s.
    After 10 seconds the valve is suddenly changed to
    position 3 (water flow 50 cm³/s) then 20 s later,
    the valve is moved suddenly to position 4 where the
    flow is 80 cm³/s. Twenty-five seconds later the
    valve is suddenly returned to position 1 where the
    flow is shut off.
    (a) Plot the flow rate against time and express this
        graph in terms of singularity functions.
    (b) Plot the volume of water that has passed through
        the valve in the time interval from $0 \le t \le 55$ s.
        Describe this volume by singularity functions.
2.7 (a) Express the singularity function

$$g(t) = 10u_p(t - 2)$$

   in the functional form (polynomial in time).
   (b) For Figure P2.7 find:
       (i)   a graph $(g_i(t))$ whose integral produces
              the figure;
       (ii)   the expression for $g_i(t)$ in terms of
              singularity functions;
       (iii) the integral of $g_i(t)$ in terms of singu-
              larity functions.

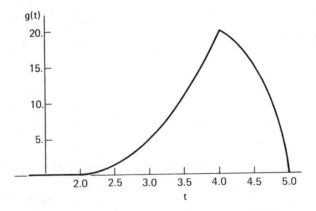

Figure P 2.7

2.8 Given: The proportional plus integral operation:

$$g_0(t) = 2g_i(t) + \int g_i(t)dt$$

and the input function:

$$g_i(t) = 5u_s(t) + 5u_s(t - 5) - 15u_s(t - 10)$$
$$+ 5u_s(t - 15)$$

   (a) Sketch the proportional part of the output.
   (b) Sketch the integral part of the output.
   (c) Sketch the complete output.

2.9 A function $g(t)$ is the product of two time functions $h_1(t)$ and $h_2(t)$. If

$$h_1(t) = u_r(t) \quad \text{and} \quad h_2(t) = u_s(t) - 2u_s(t - 5)$$
$$+ u_s(t - 10)$$

   Sketch $g(t)$.

2.10 Sketch the following time functions:
   (a) $g_1(t) = e^{-t} [I_m(e^{j2t})] u_s(t)$

   (b) $g_2(t) = e^t [Re(e^{j3t})] u_s(t)$

2.11 Figure P2.11 shows a specific function of time, $g(t)$.
   (a) Express $g(t)$ in terms of singularity functions.
   (b) Sketch the time integral and time derivative of $g(t)$.
   (c) An engineer wishes to multiply the function, $g(t)$, by another function, $h(t)$ so that the resulting function is twice as large as $g(t)$ in the interval $1 \leq t \leq 3$ and is zero elsewhere. Express $h(t)$ in terms of singularity functions.

2.12 (a) Sketch the graph of $g(t) = [\sin t] u_s(t)$.
   (b) Sketch the time integral of $g(t)$.
   (c) Find the time derivative of $g(t)$ in terms of singularity functions.

2.13 Sketch the time integral and time derivative of the following expressions:

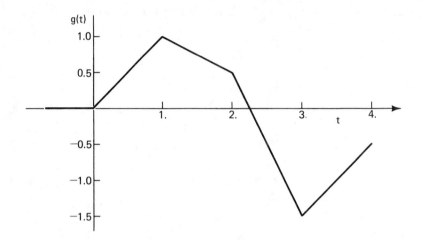

Figure P 2.11

(a) $g_1(t) = [\sin(\pi t/4)][u_S(t - 4) - u_S(t - 16)]$

(b) $g_2(t) = [\cos t][u_S(t) - 2u_S(t - \pi) + u_S(t - 2\pi)]$

2.14 Make a sketch of the following product functions:

(a) $g_1(t) = (\sin t)[u_S(t - \pi) + u_S(t - 2\pi)$
$\qquad\qquad + u_S(t - 3\pi) - 3u_S(t - 4\pi)]$

(b) $g_2(t) = [\cos t][u_S(t - \pi) + u_S(t - 2\pi)$
$\qquad\qquad + u_S(t - 3\pi) - 3u_S(t - 4\pi)]$

2.15 The relation between an input function, $g_i(t)$ and
an output function, $g_o(t)$ is given by:

$$g_o(t) = g_i(t) + 2 \int g_i(t)dt$$

If $g_i(t)$ is as shown in Figure P2.15
(a) Sketch $g_o(t)$.
(b) Express $g_o(t)$ in terms of singularity functions.

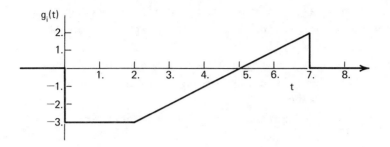

Figure P 2.15

2.16 Place the function

$$g(t) = 8 \cos 5t + 6 \sin 5t$$

in the form

(a) $g(t) = D \cos (5t - \phi)$

(b) $g(t) = D \sin (5t + \theta)$

2.17 Figure P2.17 is a function $g(t)$.

(a) Express $g(t)$ in terms of singularity functions.

(b) Sketch the time derivative of $g(t)$.

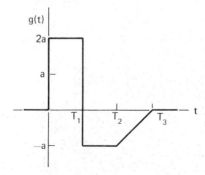

Figure P 2.17

2.18 A dispute arose between the driver of a car and a
     witness who was looking on. The car had stopped
     exactly at a wall. The driver claimed:
     (i)   135 m from the wall his velocity was 27 m/s.
     (ii)   a constant deceleration to 18 m/s in the
            first 90 m travelled.
     (iii) a deceleration of 3.6 m/s^2 thereafter.
     The witness claimed:
     (i)   81 m from the wall the velocity was 27 m/s.
     (ii) a deceleration of 3.6 m/s^2 thereafter.
     Check each claim separately. Specifically
     (a) Sketch the acceleration-time relation assuming
         at t = 0 the car is 135 m from the wall.
     (b) Express the acceleration in terms of singular-
         ity functions.
     (c) Find the velocity of the car in terms of singular-
         ity functions.
     (d) Sketch the velocity-time relation.
     (e) Which account is correct? Why?
2.19 A signal variable y(t) has the form shown in Figure
     P2.19. If the initial value of the variable at
     t = 0 (y(0)) equals zero,
     (a) Sketch the time integral of y(t).
     (b) Repeat part (a) if T_p = 0.5.
     (c) Repeat part (a) if $T_p \to 0$.

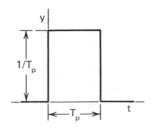

Figure P 2.19

2.20 The velocity of an automobile was measured as shown in Figure P2.20.

(a) Find the distance travelled after 8 s.

(b) Sketch the acceleration-time relation.

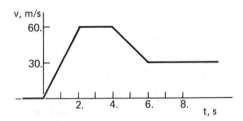

Figure P 2.20

2.21 A system is designed so that the relation between an input $g_i(t)$ and an output $g_o(t)$ is:

$$g_o(t) = 3g_i(t) + 2\frac{dg_i(t)}{dt}$$

If the input has a form shown in Figure P2.21, plot the output as a function of time.

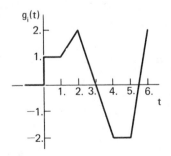

Figure P 2.21

2.22 A function may be expressed in piecewise form as:

$$g(t) = 0 \qquad\qquad , \qquad t \le 2$$
$$= 1 - \cos\pi t/2 \quad , \qquad 2 \le t \le 4$$
$$= 2(\cos\pi t/2) - 2, \quad 4 \le t \le 6$$
$$= 0 \qquad\qquad , \qquad t \le 6$$

(a) Sketch the graph of g(t)

(b) Express g(t) in terms of singularity functions.

2.23 A propellor with a radius of 0.3 m has the angular velocity ω, shown in Figure P2.23.

(a) Express ω(t) in terms of singularity functions.

(b) Find the total distance travelled by a point on the tip of the propellor after t = 70 s.

(c) Sketch the angular acceleration $\alpha(t) = [\frac{d\omega}{dt}]$ as a function of time.

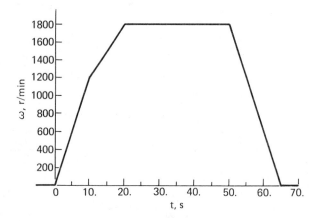

Figure P 2.23

2.24 (a) Sketch the time integral of the functions $g_1(t)$
         and $g_2(t)$ shown in Figure P2.24.
     (b) Compare the results for t greater than 2 s.

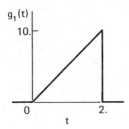
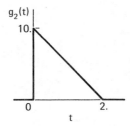

Figure P 2.24

CHAPTER 3

CIRCUIT COMPONENTS

3.1 INTRODUCTION TO CIRCUIT COMPONENTS

In this text we consider electric, fluid, mechanical and thermal circuit components. Thus we deal with four distinct physical situations.

A model of any physical situation requires the use of mathematical variables to represent certain physical quantities associated with its behaviour. There are only two basic types of such quantities (Chapter I): those which flow along or through a constrained path, such as water flowing through a pipe, and those which require, whether explicitly or implicitly, a measurement which must be made between two points such as the distance between the points.

In connecting an instrument between two points it must be
placed across everything which lies between them and such
quantities may reasonably be described by across variables.
In measuring the flow along a path it is necessary to cut
the path and insert the instrument so that the flow is
constrained to pass through it. Such quantities are
therefore described by through variables. All the media
(electric, fluid, mechanical and thermal) will have an
across variable and a through variable. The variables for
each media are different.

   The classification of all physical quantities as either
through, or across, is the basis of the recognition of
analogous situations in different physical media. This
grouping is especially useful if it is done in a manner
that ensures that the method of calculating the energy
for one medium (e.g., electric) is the same as that used
in another medium (e.g., fluid). This, however, is not a
necessary condition. The essential feature in choosing
variables is that the resulting equations for all media
shall have exactly the same mathematical form. This text
considers electrical, fluid, mechanical and thermal systems.
In the first three of these the choice of the signal variables is made so that the product of through and across
variables represents power. Thermal systems are different,
however, in that the through variable itself, the heat flow
rate, is power.

   In choosing the through and across variables it must be
kept in mind that the objective in the end is to produce
a network representation for a physical system. This
simplifies the problem of describing equilibrium conditions.
For example, the lines in Figure 3.1 may represent pipes
carrying a liquid. Equilibrium is described in part by
noting that the rate at which the liquid leaves the junction
through pipes 2 and 3 must equal the rate at which it enters
from pipe 1. This is the principle of continuity which in
this case derives from the conservation of mass. The same

thought process is used to describe the necessary conditions
which result from the conservation of charge in electrical
systems or the conservation of energy in thermal systems.

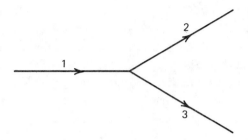

Figure 3.1 Continuity at junction

If the lines represent co-linear forces which are acting
in the same or opposite directions at a point or on a rigid
body, the equilibrium resulting from the balancing of the
forces is described in the same manner, rather more simply
than the equilibrium of a set of co-planar forces. Indeed,
the junction of connections in a network automatically
contains all the information of a free body diagram in
mechanics.

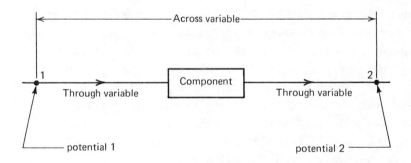

Figure 3.2 Basic two terminal component

However, before considering equilibrium in general it
is necessary to identify certain elemental situations and
to derive the elemental equations describing their performance. These correspond to idealized physical situations
in which only one physical phenomenon is considered to be
present. We model these phenomena with two terminal
elements. Figure 3.2 is a schematic representation of the
basic two-terminal component. We designate the point at
which flow enters as terminal 1 and the point at which flow
leaves as terminal 2. There is a potential associated with
each terminal which is a measure of the energy in the circuit
at that location. The across variable is the potential
difference between terminals 1 and 2. The through variable,
representing a transmitted quantity, passes through the
component without change. There are only three basic
elemental relations and they are expressed mathematically
as:

$$(\text{Across Variable}) = K_1 (\text{Through Variable}) \qquad (3.1a)$$

$$(\text{Across Variable}) = K_2 \frac{d}{dt} (\text{Through Variable}) \qquad (3.1b)$$

$$(\text{Across Variable}) = K_3 \int (\text{Through Variable})dt \qquad (3.1c)$$

where K_1, K_2, and K_3 are proportionality constants that
depend on geometric configuration and material properties.
It should be emphasized that Equation (3.1) is merely a
general form for circuit components. The basic physical
laws of electricity, fluid mechanics, mechanics, and heat
transfer can be placed in this format. It is these laws
which determine the appropriate signal variables and the
specific values of K_1, K_2, and K_3. In the following
sections we will introduce the basic elements in electrical,
fluid, mechanical and thermal systems.

3.2 ELECTRIC CIRCUIT COMPONENTS

Electric Signal Variables

Although electric charge (Q) is perhaps the most basic electrical quantity from a theoretical viewpoint, the most useful variables for electric circuits are current and voltage. Current (i) is the rate of flow of charge and is clearly identified as a through variable. The unit of charge is the coulomb (abbreviation C) and an electric current is the rate of flow of this charge in coulombs per second or amperes (abbreviated A). Thus, in mathematical terms $i = dQ/dt$.

Voltage difference, $e_1 - e_2$, (sometimes called potential difference) is the across variable. Potential difference is the work done per unit charge in moving the charge between two points at potentials e_2 and e_1. The fact that two points are required in the determination of a voltage difference (or voltage drop) clearly identifies it as an across variable. The unit for electric potential is the volt (abbreviated V). This is the potential when the work involved in moving one coulomb is one joule. Thus the work, W, done in moving a charge, Q, between two points from potentials e_2 to e_1 is $W = Q(e_1 - e_2)$. The rate at which this work is expended is the power, P. The instantaneous power is:

$$P = \frac{dW}{dt} = (e_1 - e_2)i \qquad (3.2)$$

Note that the power is the product of the across variable and the through variable.

Since the rate of change of work is defined as power this is the well-known general expression for power in an electric circuit. The unit for power is the watt (abbreviation W) which is one joule per second. It also follows that determination of energy or work is simply a matter of integrating Equation (3.2)

$$W = \int_{t_1}^{t_2} (e_1 - e_2) i \, dt \qquad (3.3)$$

There are three elemental situations which are found in circuits (Equation (3.1)), which do not involve a source of energy to be put into the system. These are commonly known as the ideal resistor, inductor, and capacitor.

Although real components only approach the ideal, they often can be modelled as a combination of two or more of these ideal components. All the components must originate from physically realizable phenomena and obey physical laws.

Electrical Resistance

From experience on many metallic circuit parts we observe that the voltage difference established in such a circuit is directly proportional to the current. This observation may be expressed in mathematical form as:

$$e_1 - e_2 = R \, i \qquad (3.4)$$

This relation is known as Ohm's Law. Note that Equation (3.4) is identical in form to the general elemental relation given in Equation (3.1a) with $R = K_1$. The quantity symbolized by R is called resistance and is a property of the material and geometry of the circuit part.

The ideal resistor has the voltage/current characteristic shown in Figure 3.3. In this case, the voltage difference is directly proportional to current as indicated by the line of constant slope. The slope of this line is the resistance, R.

The unit of resistance is the ohm (abbreviation Ω). There are occasions however, when the reciprocal, G, is more convenient. This is the conductance for which the unit is the siemens (abbreviation S).

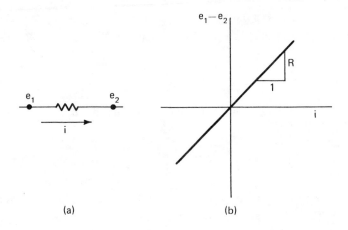

Figure 3.3 Characteristics of ideal resistance

Note that the ideal resistance element is bi-directional or bilateral. If potential 1 is greater than potential 2 then the across variable, $e_1 - e_2$, is positive and this defines the positive direction of current. When potential 2 is greater than potential 1 the across variable is negative and the current reverses. In a bilateral resistive element a given positive or negative voltage difference causes a current of the same magnitude but in opposite directions.

All objects have electrical resistance. However, in electrical circuits that are used to produce some desired effect on the signal variables we are dealing with an electric component that is especially made for the purpose. In general, resistance is obtained from the resistivity and geometry of wire material. The resistance of a wire is expressed as:

$$R = \frac{\rho \ell}{A} \ , \ \Omega \tag{3.5}$$

where ρ is the resistivity ($\Omega \cdot m$) of the wire material, ℓ is the wire length and A the wire cross-sectional area.

The energy dissipated in a resistance can be determined by rearranging Equation (3.2) in the form

$$W = \int P \, dt$$

When the voltage and current are not changing with time the energy is

$$W = R \, i^2 \, t = (e_1 - e_2)^2 t/R \qquad (3.6)$$

Note that the energy increases with time. The resistance is, therefore, an energy dissipative component.

Electrical Inductance

The most common form of inductor is a coil of wire wrapped around a core. The core contains a high level of magnetic flux due to current in the conducting wire coil. When the wire coil contains more than one loop each loop contributes to the flux linkage, λ_{12}, between the two terminals of the coil. When the physical arrangement is such that the flux-linkage is a single-valued function of current the element is called a pure inductor. This is expressed as:

$$\lambda_{12} = L \, i \qquad (3.7)$$

where λ_{12} is the flux-linkage in webers and L is the inductance in henrys (abbreviated H). Thus the pure or ideal inductor has the flux-linkage/current characteristic shown in Figure 3.4. Faraday's Law states that the voltage difference induced in coils by a changing magnetic field is

$$e_1 - e_2 = \frac{d \lambda_{12}}{dt} \qquad (3.8)$$

The combination of Equations (3.7) and (3.8) yields a relation between the signal variables, that is:

$$e_1 - e_2 = L \frac{di}{dt} \qquad (3.9)$$

Equation (3.9) has the same form as given in the general elemental relation in Equation (3.1b). In this case $K_2 = L$. From the ideal characteristics shown in Figure 3.4, we may observe that the ideal inductor is also a bilateral component.

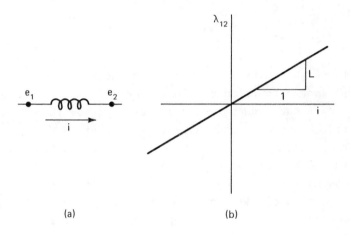

(a)   (b)

Figure 3.4 Characteristics of ideal inductance

The magnitude of inductance for a wire wound over a cylindrical core is:

$$L = \frac{4\pi (10)^{-7} \; \mu \; A \; N^2}{\ell} \;, \; H \qquad (3.10)$$

where μ is the relative permeability of the core material, A is the area of the core (m^2), ℓ is the length of the core (m), and N is the number of coils. For an air core the relative permeability $\mu = 1$. When an iron rod is used for the core the relative permeability may be as high as 10 000, the actual value depending on the flux density.

In actual electrical elements it is not possible to achieve the ideal characteristics given for the resistance

and inductance.   In both cases the element consists of a
length of wire.   Thus the actual inductance always has
some resistance and the actual resistance has some
inductance.   However, we will assume throughout the text
that the components are ideal and represent only one
phenomenon.

The energy into an inductor is from Equations (3.2) and
(3.9),

$$W = L \int_{i_0}^{i} i' \, di' = \frac{L}{2} (i^2 - i_0^2) \qquad (3.11)$$

where i_0 is the initial current.   For the inductor the
energy is not a function of time directly.   The reason for
this is that the inductor stores energy rather than dissipates it.   The energy stored by the inductor is due to the
flow of current and the expression given in Equation (3.11)
is valid whether or not the current is constant.   The energy
stored is not directly dependent on the voltage difference.
Indeed Equation (3.11) applies even when the voltage difference is zero.   The inductor is in a class of energy storage
elements known as T-type (for through variable) storage.

Electrical Capacitance

Two conductors separated from each other by an insulating
material (called dielectric) form a capacitor.   In the
capacitor an electric field is established between the
conductors.   When the physical arrangement is such that the
charge on the conductors is a single-valued function of the
voltage difference, the element is called a pure capacitor.
This is expressed as:

$$Q = C(e_1 - e_2) \qquad (3.12)$$

where Q is the charge in coulombs and C is the capacitance
in farads (abbreviated F).   Thus the pure or ideal capacitor
has the charge/voltage difference characteristic shown in

Figure 3.5. In terms of the signal variables we may replace
the charge Q in Equation (3.12) by the integral of the
current so that:

$$e_1 - e_2 = \frac{1}{C} \int i \; dt \qquad\qquad (3.13)$$

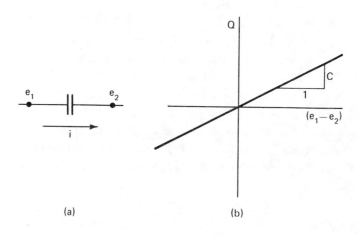

(a)   (b)

Figure 3.5 Characteristics of ideal capacitance

For this type of component the governing equation (Equation
(3.13)) has the same mathematical form as the general elemental relation of Equation (3.1c) with $K_3 = \frac{1}{C}$. The ideal
capacitor is a bilateral component in the same way as the
ideal resistor and ideal inductor.

For a parallel plate capacitor the magnitude of the
capacitance is:

$$C = \frac{\varepsilon A}{d} \; , \; F \qquad\qquad (3.14)$$

where ε is the permittivity of the dielectric in farads
per metre, A is the cross-sectional area (m^2) of the plates,
d is the thickness (m) of the dielectric.

The energy associated with a capacitance may be determined
from Equations (3.2) and (3.13) as:

$$W = C \int_{\Delta e_0}^{\Delta e} e'de' = \frac{C}{2} (\Delta e^2 - \Delta e_0{}^2) \qquad (3.15)$$

where Δe is the voltage difference, $e_1 - e_2$ and Δe_0 is the initial voltage difference. Since a capacitor stores energy it is also an energy storage component. In this case, energy is stored due to the presence of a voltage difference. The expression for energy given in Equation (3.15) is valid whether or not the voltage difference is constant. The energy stored is not directly dependent on the current. Indeed Equation (3.15) applies even when the current is zero. The capacitor is in the class of energy storage devices known as A-type (for across variable) storage.

The energy stored by an inductance (Equation (3.11)) and by a capacitance (Equation (3.15)) may be formulated in a similar way for the analogous storage elements in fluid and mechanical systems.

The normal symbols for the three circuit elements are shown in Figures 3.3, 3.4, and 3.5. The arrow associated with the current shows the positive direction of conventional current flow through the element. The terminal at which positive current enters is positive for each of the elements and polarity indicators must be set up as in the figures if the positive signs implied in the elemental equations are to apply.

At this stage, it can be noted that each of the three elements is significantly different. One of them, the resistor, dissipates energy which cannot be recovered by the circuit. The others store energy which can be recovered when the circuit conditions are suitable. However, in one case, the capacitor (A-type storage) the energy is stored by virtue of there being a voltage between the two terminals irrespective of whether or not there is current flow. In the other case, the inductor (T-type storage) the opposite statement holds: the energy is stored by virtue of there being a current flow through the element irrespective of whether or not there is a voltage across it.

In this chapter, the circuit models of the storage elements do not include initial storage. That is, i_0, for the inductor and Δe_0, for the capacitor have been assumed to be zero. The analogous fluid, mechanical and thermal storage elements will also be introduced without initial storage. We will consider the circuit models for storage elements with initial storage in Chapter IV.

3.3 FLUID CIRCUIT COMPONENTS

To develop models for components with fluid flow we must apply the equations of fluid mechanics to these components and try to adapt them to the forms given for the basic elements in Equation (3.1). In some cases, the components analyzed by the fluid mechanics equations do not match the basic forms exactly. When that happens, restrictions are placed on the resulting fluid components. Let us be aware, therefore, that the analogy between fluid components and electric components is not perfect.

Fluid Signal Variables

The quantity that is conserved in the flow of fluid at a junction is the mass flow, m, in kg/s. Strictly speaking this ought to be the through variable. However, it is more convenient for measurement purposes to use the volume flow, q, in m^3/s as the through variable. The relation between mass flow and volume flow is $m = \rho q$ where ρ is the fluid density in kg/m^3. Thus the volume flow will be proportional to the mass flow when the fluid density remains constant throughout the circuit. Under this condition, the volume flow becomes an exact through variable. The constant density condition is approximated very closely when we have a gas medium with low velocity or a liquid medium. We therefore restrict the fluid components to liquid flow or incompressible gas flow when we select volume flow as the through variable.

The choice of a suitable across variable is suggested by
the condition that the product of through and across variables should represent power.

From this we determine that the units of an appropriate
across variable are N/m^2 or pascals (abbreviated Pa). Thus
the pressure difference, $p_1 - p_2$, is clearly indicated as
the across variable in fluid systems.

One interesting result is that it is possible to define
pressure in a manner which is directly analogous to the
definition of voltage. That is, the difference in pressure,
Δp, is the work done, ΔW, in moving unit volume, ΔV, between
two points. This is illustrated by considering the movement
of an incremental volume of fluid between two points along
a pipe of uniform cross-section A, as shown in Figure 3.6.
The work done according to this alternate definition is
given by

$$\Delta p = \frac{\Delta W}{\Delta V} \qquad\qquad (3.16)$$

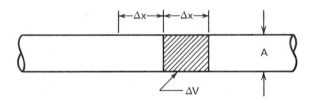

Figure 3.6 Elemental volume within a pipe

However, the volume may be expressed as $\Delta V = A\Delta x$, and
the work done is the net force on the incremental volume
multiplied by the distance moved. That is $\Delta W = (\Delta F)(\Delta x)$.
The substitution of the work and volume into Equation (3.16)
yields:

$$\Delta p = \frac{(\Delta F)(\Delta x)}{A(\Delta x)} = \frac{\Delta F}{A} \qquad\qquad (3.17)$$

In the limit, this is the force per unit area which is the
normal definition of pressure.

Just as there are three elemental physical situations
with electric circuits, there are three directly analogous
situations in problems of the flow of fluids. However,
depending on the fluid involved the linear relationships
between the through and across variables do not exist over
as wide a range of values and the resulting models are
therefore not quite as accurate. Nevertheless, a significant amount of information can be derived from such models
leading to considerable insight into how fluid systems
behave. Since the amount of effort involved in obtaining
solutions for linear systems is significantly less than
that required for non-linear systems, the linear model
is an attractive tool for determining the general nature of
the performance of fluid systems. It is of course, necessary to have some appreciation of the conditions under
which the linear model ceases to be accurate, and part of
the following discussion is directed towards making this
assessment.

Fluid Resistance

The fluid resistance models frictional effects which
impede the flow of fluid along a pipe. These exist at the
interface between the fluid and the pipe, and also between
layers of fluid which are not moving at the same velocity.
In certain circumstances, the flow of all the particles of
fluid is in the direction of the tube such as the one indicated in Figure 3.7, this form of flow being termed laminar.
The profiles of fluid velocity are shown at three locations
in the pipe. At the pipe entrance the velocity profile is
uniform. The viscous properties of the fluid cause shear
forces which act to retard the velocity at the boundaries
and increase it along the centerline. At some distance down
the pipe, a stabilized parabolic shape profile is reached.
This distance is known as the entrance length, ℓ_e, and

usually varies from about 80 to 120 diameters. Thereafter,
the velocity profile remains unchanged. In the region where
the profile is not changing, the steady state momentum
equation of fluid mechanics yields:

$$P_1 - P_2 = Rq \qquad\qquad (3.18a)$$

$$R = \frac{128\ \mu(\ell - \ell_e)}{\pi d^4} \qquad\qquad (3.18b)$$

where μ is the dynamic viscosity of the fluid in Pa·s
      d is the diameter of the pipe in m.

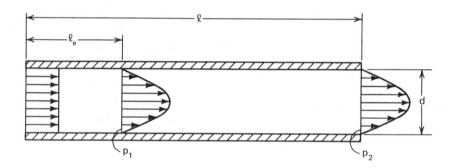

Figure 3.7 Pipe flow with viscous friction

Thus, for laminar flow only, the pressure drop along a
pipe is proportional to the flow rate and the fluid resistance is defined in exactly the same manner as the
electrical resistance. The linear portion of the pipe
resistance occurs only after the entrance length. Since
there is no way to eliminate the non-linear entrance effects
the usual procedure is to select pipe lengths, ℓ, that are
at least four times the entrance length. In this way,
entrance length effects are minimized. Actually, the fluid
resistance is usually approximated by replacing the distance,
ℓ - ℓ_e, in Equation (3.18b) with the length, ℓ, alone. Thus,
as a good approximation when, ℓ, is much larger than, ℓ_e

$$P_1 - P_2 = Rq \qquad (3.19a)$$

$$R = \frac{128 \; \mu \ell}{\pi d^4} \qquad (3.19b)$$

where p_1 is now the pressure at the entrance to the pipe.
We should emphasize that Equation (3.19) holds only for
laminar flow.

When the flow rate increases beyond some critical value,
the flow breaks up and contains circulating currents called
vortices. This type of flow is called turbulent and results
in an increased rate at which energy is dissipated within
the fluid. The critical point between laminar and turbulent
flow is not precisely known but may be estimated by means
of the Reynolds number (Re) which is a non-dimensional
expression involving several properties of the fluid and
the geometry of the pipe.

$$Re = \frac{4}{\pi} \frac{\rho}{\mu} \frac{q}{d} \qquad (3.20)$$

Normally, if the Reynolds number is less than 2 000, the
flow is laminar, but becomes fully turbulent when the
Reynolds number is 5 000.

No special name has been given to the unit of fluid
resistance and probably the most convenient unit is $N \cdot s/m^5$.
The laminar flow fluid resistance given in Equation (3.19b)
is applicable for both liquid and gas flows. We will limit
our consideration to two common fluid media, water and air.
At 20°C the properties of water and air are:

	Water	Air
Viscosity	$\mu = 10.1(10)^{-4} Pa \cdot s$	$\mu = 0.18(10)^{-4} Pa \cdot s$
Density	$\rho = 1\ 000 \ kg/m^3$	$\rho = 1.2 \ kg/m^3$

A fluid resistance is bilateral if a flow in either
direction develops the same magnitude of pressure difference
between the end terminals. In fluid systems we may assume

at once that the resistances are bilateral if only one fluid
medium is present.   However, in cases where liquid pipes
empty into air the resistance is not bilateral.   Consider,
for example, Figure 3.8.   When the resistance is connected
between the pump and tank (Figure 3.8a) it is bilateral.
There is only one medium, the water.   Now on the other hand,
when the resistance is connected between water pump and air
(Figure 3.8b) the resistance is not bilateral.   A reversal
of the pump will draw in air.   We may still apply our circuit
model, however, if we know in advance that the flow always
remains in the proper direction from pump to atmosphere.

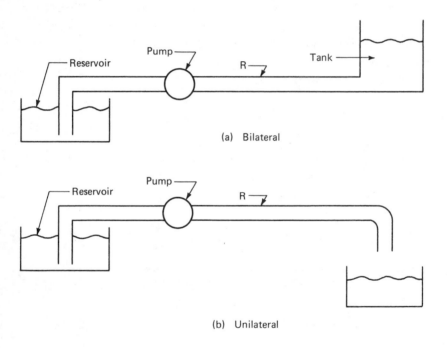

Figure 3.8 Fluid resistances

  The power dissipated by a fluid resistance is equal to
the product of the pressure drop and flow rate associated
with the component.   This may also be expressed as Rq^2 or
$(p_1 - p_2)^2/R$.

Fluid Inertance

Another physical situation in fluid systems relates to
the inertia effect in the fluid. This is of particular
importance with liquids although it is not necessarily
negligible in pneumatic systems. Consider a tube as indicated in Figure 3.9, in which the fluid volume flow rate is
q m^3/s. If we restrict our attention to a section of length
ℓ and uniform cross-sectional area A, the mass of the fluid
is $\rho A \ell$. The velocity of the fluid is q/A and hence the
momentum is $\rho \ell q$. From the momentum equation of fluid
mechanics for frictionless flow the rate of change of
momentum is equal to that of the net force acting on the
mass of fluid. This net force is equal to the product of
pressure difference and area. Thus

$$P_1 - P_2 = \frac{\rho \ell}{A} \frac{dq}{dt} \qquad (3.21)$$

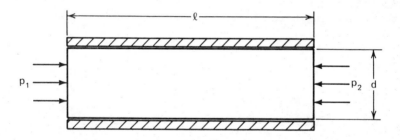

Figure 3.9 Fluid inertance model

Equation (3.21) has the same form as the general relation
given in Equation (3.1b). Pressure difference, $P_1 - P_2$, as
the across variable is directly analogous to voltage difference and q is directly analogous to current. We therefore note that this equation has identically the same form
as the elemental equation for inductance (Equation (3.9)).
This effect can therefore be called fluid inductance
although usually the term inertance is used, since it is
the inertia effect which is being modelled.

The fluid inertance of a pipe is calculated from

$$L = \frac{\rho \ell}{A} \qquad (3.22)$$

The unit for fluid inertance is $N \cdot s^2/m^5$.

The energy stored by a fluid inertance is dependent upon the flow rate (through variable) and inertance in the same form as given for the electrical inductance in Equation (3.11).

Fluid inertance has been derived under the assumption of the flow of a frictionless fluid in a pipe. Previously fluid resistance was also derived in a pipe for a fluid with friction but without inertia. Both components share the same physical geometry: the pipe. For analysis purposes, it is accepted to separate the friction and inertia effects in the pipe. In any practical case, however, both effects exist simultaneously. Thus, in a real pipe there is both fluid resistance and fluid inertance. There is only one flow, q, passing through the pipe. The complete model for a pipe is therefore:

$$P_1 - P_2 = Rq + L \frac{dq}{dt} \qquad (3.23)$$

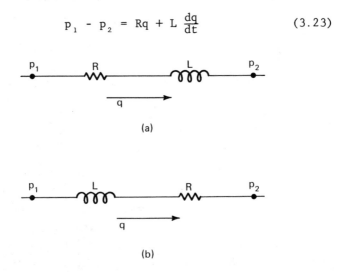

(a)

(b)

Figure 3.10   Fluid line circuit models

The pressure difference, $p_1 - p_2$ is the sum of the pressure differences caused by fluid resistance and fluid inertance. This model is shown in Figure 3.10. We use the same symbols for the fluid components that we previously used for the electrical components. The model of a fluid line or fluid pipe consists of a resistance and inertance in series. The order of the components does not matter for the fluid line because there is no definable pressure terminal between the resistance and inertance. This occurs because the effects are not separable in a physical sense although we did separate them analytically. Sometimes one of the effects is much smaller than the other and can be neglected. We will indicate in each problem the appropriate model for each line component. However, in general, both effects must be considered.

Fluid Capacitance

The momentum equation of fluid mechanics was used to define fluid resistance and inertance. Fluid capacitance is due to an effect we may observe in the continuity equation of fluid mechanics. A simple way to demonstrate the formulation of the signal variables in the form of Equation (3.1c) is to consider the closed chamber (Figure 3.11a). The mass of fluid, M, in the chamber is equal to the product of density and volume ($M = \rho V$). If we take the time derivative of the mass we obtain an expression for the mass flow, m. In terms of the volume flow the result is:

$$q = \frac{dV}{dt} + \frac{V}{\rho} \frac{d\rho}{dt} \qquad (3.24)$$

Equation (3.24) is a form of the continuity equation of fluid mechanics. When the volume of the chamber is fixed the first term on the right-hand side is equal to zero. The resulting equation relates volume flow to fluid density. However, our signal variables are volume flow and pressure difference. To find a relation between the signal variables we must specify the connection between pressure and

density. This connection depends on the fluid. For a
liquid we may apply Hooke's Law to obtain:

$$\frac{d\rho}{\rho} = \frac{dp}{\beta} \qquad (3.25)$$

where β is the bulk modulus of the liquid in Pa. For water
$\beta = 2.14(10)^9$ Pa. When the fluid is a gas the relation
between pressure and density depends on the thermodynamic
process. The appropriate results are:

$$\frac{d\rho}{\rho} = \frac{dp}{(p + p_a)} \quad \text{(isothermal process)} \qquad (3.26a)$$

$$\frac{d\rho}{\rho} = \frac{dp}{\gamma(p + p_a)} \quad \text{(adiabatic process)} \qquad (3.26b)$$

where p_a is the atmospheric pressure and γ is the ratio of
specific heats. We emphasize the atmospheric pressure here
because the absolute pressure is required in Equation (3.26)
and p is usually the gage pressure. If Equations (3.25) and
(3.26) are substituted into Equation (3.24) and an integration performed

$$p = \frac{\beta}{V} \int q \, dt + p(0) \qquad \text{(liquid)} \qquad (3.27a)$$

$$p = \frac{(p + p_a)}{V} \int q \, dt + p(0) \quad \text{(gas, isothermal)} \qquad (3.27b)$$

$$p = \frac{\gamma(p + p_a)}{V} \int q \, dt + p(0) \quad \text{(gas, adiabatic)} \qquad (3.27c)$$

where p(0) is the value of the pressure in the chamber at
t = 0.

If we assume that p(0) = 0, Equation (3.27) has almost
the desired form given in Equation (3.1c). The difference
between Equation (3.27) and Equation (3.1c) is that there
is only one pressure in Equation (3.27) and a pressure
difference is required to fit the standard form. This is
a consequence of the fact that the closed tank has only

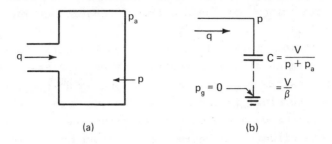

Figure 3.11 Closed tank capacitance

one terminal which is the internal pressure, p. In effect, then, we are trying to make a two-terminal component model out of a physical component that has only one terminal. We accomplish this by assuming a fictitious terminal, p_g, at zero potential and expressing Equation (3.27) as:

$$p - p_g = \frac{1}{C} \int q \ dt \qquad (3.28)$$

where $C = \frac{V}{\beta}$ for a liquid and

$$C = \frac{V}{(p + p_a)} \quad \text{or} \quad \frac{V}{\gamma(p + p_a)}$$

for a gas. The unit of fluid capacitance is m^5/N.

Thus capacitance for a gas is a function of pressure, p. When the system operates with small pressure changes, the quantity $p + p_a$ does not change appreciably and the capacitance is essentially constant. When there are large pressure transients the capacitance value is variable. In general, a component that varies as a function of one of the signal variables is called a non-linear component. Such components are beyond the scope of this text. For our purposes all problems relating to gas operation will be restricted to small pressure transients. Then we can treat the capacitance as a constant.

The energy stored by a fluid capacitance is dependent on the pressure difference (across variable) and has the same

form as that given for the electrical capacitance in
Equation (3.15).

The fictitious terminal adopted to complete the two-
terminal form is at zero potential. Thus, a closed tank
always represents a capacitance to ground. However, since
the ground terminal is fictitious, it cannot be located
physically. We denote a fictitious ground terminal by
dashing the circuit line going to the ground. Figure 3.11b
shows the circuit representation for a closed tank capacitor. Notice that the liquid filled closed tank is not
bilateral. If liquid is removed from the tank, cavitation
occurs and we are no longer dealing with a homogeneous
medium. A gas filled tank is bilateral up to a point. It
ceases to be bilateral when all the gas has been removed.

Figure 3.12a shows a fixed volume tank with inlet and
outlet flow. The previous reasoning is still applicable.
There is only one pressure terminal in the tank and our
model (Equation (3.1)) requires two pressure terminals for
each component. Analysis of the tank with inlet and outlet
flows q_1 and q_2 leads to:

$$p - p_g = \frac{1}{C} \int (q_1 - q_2)dt \qquad (3.29)$$

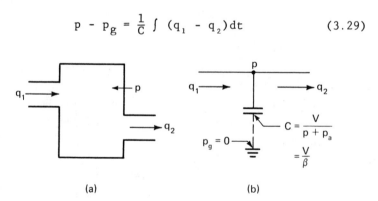

(a)   (b)

Figure 3.12 Closed tank capacitance with inlet and outlet

The circuit equivalent for this tank is shown in Figure
3.12b. The tank again represents a capacitance to ground.

Consider now the open tank in which a liquid is stored, as indicated in Figure 3.13. In this case, the pressure p is indicated at the bottom of the tank. The pressure is the gravitational force per unit area of the column of liquid above it. If the horizontal area of the tank is A, then the weight of the column of height h is

$$w = (p - p_g)A = \rho gAh$$

from which

$$p - p_g = \rho gh \qquad (3.30)$$

where g is the gravitational constant (= 9.81 m/s^2) and p_g is the pressure above the liquid column.

Before proceeding to identify the capacitive effect it may be noted that Equation (3.30) provides the justification for using the height of the column of liquid as an

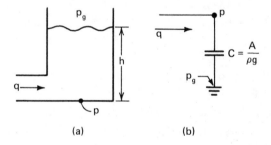

Figure 3.13 Liquid open tank capacitance

alternate measure of pressure, commonly called head. However, from the discussion at the beginning of this section, it should be evident that the product of head and flow rate cannot possibly result in power as expressed in any acceptable unit. The resulting confusion with expressions involving fluid resistance, capacitance and inductance is such that head is not used as an across variable in the fluid circuits analyzed in this text.

Now, from Equation (3.24) when density is not changing, but volume is changing q = dV/dt. In addition, since V = Ah, we may express Equation (3.30) in terms of pressure difference and flow rate as:

$$(p - p_g) = \frac{\rho g}{A} \int q \, dt \qquad (3.31)$$

Comparison with the standard form in Equation (3.1c) shows that the capacitance of the open tank is

$$C = \frac{A}{\rho g} \qquad (3.32)$$

This component is essentially a two-terminal component without the assumption of a fictitious terminal. One terminal exists at the bottom of the tank. The other is the pressure over the liquid column. Usually, this pressure will be zero gage pressure and the open tank like the closed tank will represent a capacitance to ground (Figure 3.13b). However, this is not necessarily the case. Consider Figure 3.14a, in which a closed tank is partially filled with liquid. Another way of viewing this is that we put a cover on the open tank. In the partially filled tank, the addition of flow, q, compresses the gas trapped above the liquid column. As a result, the pressure above the liquid column is no longer constant. The equivalent circuit for this circumstance is shown in Figure 3.14b. From the liquid circuit point of view, there are two capacitances in series. The resulting capacitance is reduced. Thus, sealing an open tank will reduce the capacitance.

The liquid open tank capacitance is bilateral only if there is liquid in the tank. When there are flow reversals, the liquid level in the tank must be checked to ensure that the tank has not emptied.

There is still another type of fluid capacitance. This type features a moving part between two chambers. Figure 3.15a shows a spring loaded piston separating two chambers

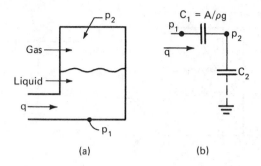

Figure 3.14 Liquid tank capacitance closed at top

at pressures p_1 and p_2. The change in volume depends on the pressure difference and spring constant, k. If the area of the chamber is A the volume change may be expressed as:

$$\Delta V = \frac{A^2}{k} (p_1 - p_2) \qquad (3.33)$$

and from Equation (3.24) (for no density change) we may relate pressure difference and volume flow for this component as:

$$P_1 - P_2 = \frac{k}{A^2} \int q \, dt \qquad (3.34)$$

Now, once again, comparison with the standard form (Equation (3.1c)) shows that this is a two-terminal capacitance and has a capacitance value of

$$C = \frac{A^2}{k} \qquad (3.35)$$

Figure 3.15b shows the equivalent circuit. The capacitance calculated above is designated as C_2. There is a compressibility capacitance associated with each chamber These are C_1 and C_3. Most often, the capacitances C_1 and C_3 will have small values. In that case, the spring-loaded piston represents a two-terminal fluid capacitance.

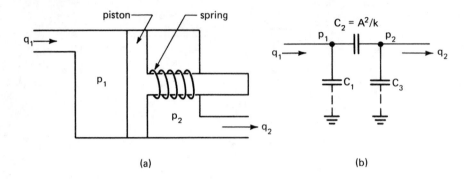

Figure 3.15 Spring-loaded piston type capacitance

3.4 MECHANICAL CIRCUIT COMPONENTS

The procedure for determining circuit models for mechanical components is similar to that used for fluid components. In this case, we apply Newton's Laws to mechanical
systems and then try to fit the resulting equations into
the standard two-terminal forms. Here again, the equations
do not always fit the standard forms exactly, and we must
use fictitious terminals.

Mechanical Signal Variables

The key to the network representation of mechanical
systems is the recognition of force as the through variable
for mechanics of translation and torque for mechanics of
rotation. Here, we note that it is quite common to speak
of forces being transmitted through certain parts or of
torques being transmitted along shafts. This transmission
does not require any motion. In the case of rotation the
torque is transmitted in the axial direction and is not
directly associated with the rotary motion. Also, there
is no quantity of fluid or charge which is being transmitted
and in this sense, the mechanical system is rather different
from the electrical and fluid systems considered so far.
Nevertheless, the concept of a "flow" of force or torque

is a great help in providing a systematic view of mechanical systems.

Consider the blocks A, B, C, (Figure 3.16a) in equilibrium on a smooth table. A horizontal force, f, is applied to block A, as shown. To demonstrate the transmissibility of force, we draw free body diagrams of blocks, A, B, and C, in Figure 3.16b (vertical forces have been neglected for clarity). Since the blocks are in equilibrium, there must be no unbalanced force acting on them. The applied force f acting to the right on block A, must therefore be balanced by a force of the same magnitude acting to the left on block A. This latter force is applied to block A by block B. From Newton's third law, every action has an equal and opposite reaction. Since block B pushes against block A, there is a reaction of block A against block B. Thus, on the free body diagram of block B, the force from block A acts to the left. To maintain block B in equilibrium requires a force from block C. The process of action and reaction shows that the force is transmitted through the blocks to the wall.

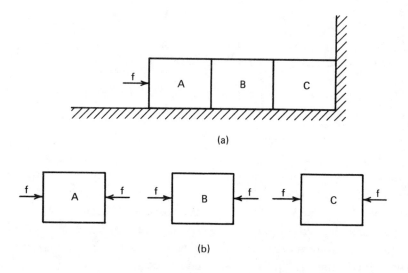

(a)

(b)

Figure 3.16 Transmission of force

In a similar way, we may demonstrate the through variable
nature of torque.   Figure 3.17a shows three disks, A, B,
and C connected together by dowel pins.   A torque, T, is
applied to disk A.   The free body diagrams of each disk
(Figure 3.17b), show that the torque T = fd is transmitted
through the disks to the wall.   The balancing torques on
each disk come from consecutive application of the action
and reaction principle.

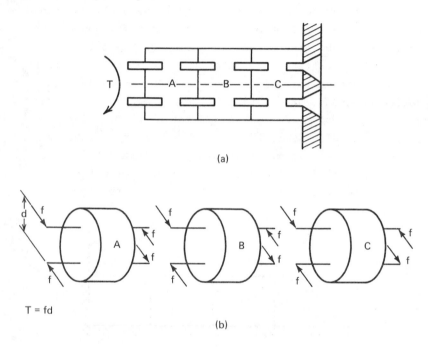

(a)

T = fd

(b)

Figure 3.17 Transmission of Torque

The across variable in mechanical systems can be distance
(or displacement) or velocity or acceleration, the choice
being made to ensure that the three elemental physical
situations are modelled by elemental equations of identically the same mathematical form as those in the electric and
fluid circuits, and that the product of the through and
across variables is power.

The appropriate across variable is therefore velocity difference and the dimensional analysis of the product is shown for confirmation

$$[\Delta v][f] = \frac{m}{s} \cdot N = \frac{N \cdot m}{s} = \frac{J}{s} = W$$

Thus, the product of velocity in m/s and force in N gives the power directly in watts. For rotation, the product of angular velocity difference ($\Delta\omega$) in rad/s and torque in N·m is also in watts.

Damping (Mechanical Resistance)

The elemental mechanical situation in which energy is dissipated is that involving friction. Unfortunately, friction effects are varied, although generally, three basic forms are recognized:

  a) Static

  b) Coulomb

  c) Viscous

Static friction is the effect whereby the force required to move a stationary object is greater than that required to move the same object when it is moving very slowly. Clearly this does not conform to the linear relationship which is analogous to the electrical resistance. Similarly, Coulomb friction, (that is force independent of velocity) does not conform to the linear situation in which force is proportional to velocity and will be considered in Section 3.6. However, there are frictional effects and devices in which the force is proportional to velocity, this being termed viscous friction. These devices are usually called dampers, although the most common example is probably the shock absorber on an automobile.

A schematic drawing of a translational damper is shown in Figure 3.18a. The damper has a piston-cylinder arrangement. However, the damping effect is due to a viscous liquid (oil) in the cylinder. The piston and cylinder

may move at different velocities. Thus, the damper is a
two-terminal device. One terminal at piston velocity and
the other terminal at cylinder velocity. The shear force
developed in the damper is proportional to the difference
in velocities of piston and cylinder. From the free body
diagram of Figure 3.18b, we find that

$$f = A\tau = \frac{(\pi Dz)\mu}{h} (v_1 - v_2) \qquad (3.36)$$

where D is the piston diameter, z is the piston length,
h is the piston clearance, τ is the shear stress, and μ
is the viscosity of the oil.

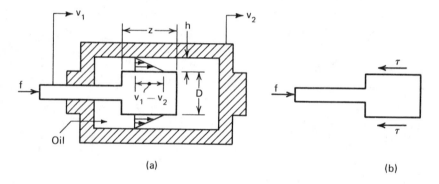

(a)   (b)

Figure 3.18 Schematic of translational damper

The usual way of defining the damping, b is to express
this proportional response as $f = b(v_1 - v_2)$.

Consideration of the analogous electric circuit variables
shows that the damping, b is therefore analogous to the
electrical conductance, 1/R. From Equations (3.1a) and
(3.36) we may describe damping as:

$$b = \frac{1}{R} = \frac{\pi Dz\mu}{h} \qquad (3.37)$$

As with the fluid resistance, there is no special unit
for damping which is most conveniently expressed in N·s/m
in translational systems.

In rotational systems, the rotational damping, B, is defined by $T = B(\omega_1 - \omega_2)$ where $\omega_1 - \omega_2$ is the angular velocity difference. The rotational damper has essentially the same geometry as the translational damper. The rotational damping is approximately

$$B = \frac{1}{R} = \frac{\pi D^2 z \mu}{2h} \qquad (3.38)$$

The unit of rotational damping is N·m·s.

Figure 3.19 shows the symbols used for damping in translational and rotational systems. The symbols on the left are descriptive indications of the process. As a circuit component, however, both dampers are represented by the symbol for resistance. Remember, however, that B is analogous to conductance (1/R).

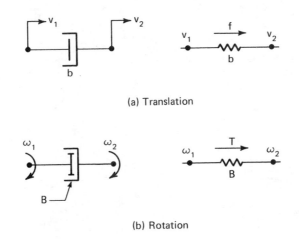

(a) Translation

(b) Rotation

Figure 3.19 Symbols for damping

In general, frictional effects do not conform exactly to any one of the three friction models. However, other than the further consideration of Coulomb friction (Section 3.6), at this stage, it is quite reasonable to approximate the actual friction characteristic with the nearest equivalent (viscous) damping. When this is done, the analytical techniques used to predict system behavior can then be exactly the same as those used in electric circuits.

The power dissipated by a damper is equal to the product of the across and through variables associated with the component.

Spring (Mechanical Inductance)

Another mechanical situation is that of a spring (Figure 3.20). The ideal spring (negligible mass) is a direct application of Hooke's law; the extension or compression is directly proportional to the force producing it. Normally, this is expressed in the form

$$f = k(x_1 + \Delta x_1 - x_2 - \Delta x_2) \qquad (3.39)$$

where k is the spring stiffness of the translational spring and x_1 and x_2 are the positions of the ends of the spring relative to some fixed reference when the force is zero. Δx_1 and Δx_2 are the displacements of the ends when the force is applied.

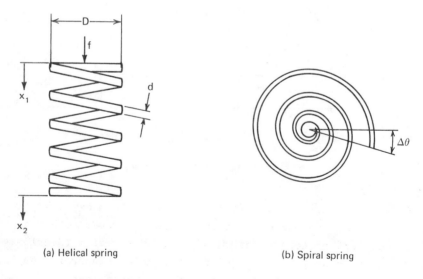

(a) Helical spring   (b) Spiral spring

Figure 3.20 Typical spring configurations

If we differentiate Equation (3.39) with respect to time and since x_1 and x_2 are constants, we obtain

$$\frac{df}{dt} = k \frac{d}{dt} (\Delta x_1 - \Delta x_2) = k(v_1 - v_2) \qquad (3.40)$$

This is in the same form as Equation (3.1b). Thus, a spring is the analogous situation to an inductor, although it must be noted that the stiffness, k, is analogous to the reciprocal of inductance, 1/L. The basic unit of stiffness is N/m which is rather small for most useful springs and therefore k is more likely to be expressed in kN/m or MN/m. For a helical spring, the spring constant may be calculated from the formula

$$k = \frac{1}{L} = \frac{Gd^4}{8ND^3} \qquad (3.41)$$

where G is the shear modulus of elasticity, d is the wire diameter, D is the coil diameter, and N is the number of coils.

For a torsional spring (Figure 3.20b), the torque is related to the change in angular displacement $\Delta\theta$, as:

$$T = K(\Delta\theta) \qquad (3.42)$$

Thus the spring constant for the torsional spring has the same units as torque (N·m). Now, the derivative of Equation (3.42) with respect to time yields

$$\frac{dT}{dt} = K(\omega_1 - \omega_2) \qquad (3.43)$$

where ω_1 and ω_2 are the angular velocities of the inner and outer terminals of the spiral spring. Note, also, that the helical spring may also be used in torsion. The spring constant for a torsional spring with round wire is:

$$K = \frac{1}{L} = \frac{E\pi d^4}{64\ell} \qquad (3.44)$$

where E is the modulus of elasticity, d is the wire diameter,
and ℓ is the total length of the spring.

Figure 3.21 shows the mechanical and circuit symbols
for springs. The circuit representation for both types
is the same as the electric inductance and the fluid
inertance. Note, however, that K is analogous to 1/L.

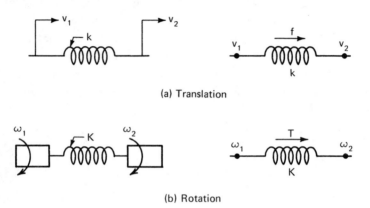

(a) Translation

(b) Rotation

Figure 3.21 Symbols for springs

The energy stored by a spring depends on the through
variable and the spring constant. It has the same form
as given for the energy stored in the electrical inductance
(Equation (3.11)).

The spring produces a force proportional to displacement.
For some special geometries a force proportional to displacement can be obtained without a spring. Consider the
harmonic motion shown in Figure 3.22. When a mass is
connected to a spring (Figure 3.22a), the system will
oscillate if the mass is initially displaced and then
released. The spring creates a restoring force that is
always proportional to the displacement. A similar restoring force may occur due to position without a spring
present. Figure 3.22b shows a mass suspended from a
string in the form of a pendulum. Now, if the mass is

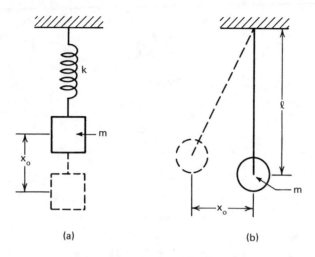

Figure 3.22 Harmonic motion

displaced and released, the mass will oscillate. The position of the mass is providing the restoring force in this case. This force can be calculated by considering a free body diagram of the mass (Figure 3.23a). From a summation of forces (Figure 3.23b) the relation between restoring force and displacement is:

$$f_R = (\frac{mg}{\ell})x \qquad (3.45)$$

where ℓ is the length of string in meters, and m is the mass in kilograms.

If Equation (3.45) is differentiated with respect to time, the result is:

$$\frac{df_R}{dt} = \frac{mg}{\ell} v \qquad (3.46)$$

The restoring force and the velocity have the relationship indicated by Equation (3.1b) except that here again, we have only one terminal. We must add a fictitious terminal

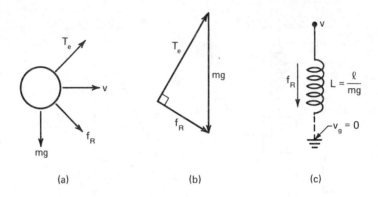

Figure 3.23 Pendulum restoring force

to make this fit into the two-terminal model. In this
case, we choose the ground at zero velocity as the other
terminal. Figure 3.23c shows the circuit representation
for restoring force due to pendulum position. The component is shown dashed because of the assumption of a
fictitious terminal. The equivalent spring constant for
the pendulum is:

$$k = \frac{1}{L} = \frac{mg}{\ell} \qquad (3.47)$$

Mass (Mechanical Capacitance)

The elemental situation of an isolated mass (Figure
3.24a) can be considered, as a simple application of
Newton's second law:

$$f = m\frac{dv}{dt} \qquad (3.48)$$

or integrating with respect to time

$$v = \frac{1}{m} \int f\, dt + v(0) \qquad (3.49)$$

When the initial velocity $v(0)$ is zero, Equation (3.49)
almost fits the model described in Equation (3.1c). The
difference is that the mass is not a two-terminal device.

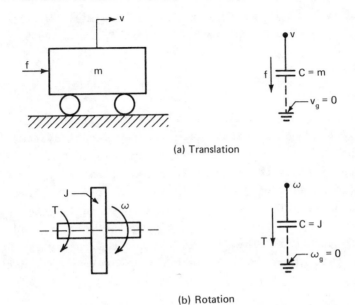

(a) Translation

(b) Rotation

Figure 3.24 Symbols for mass

There is only one velocity associated with the mass. To place this in a component form, we must add the ground terminal at zero velocity to Equation (3.49). Then Equation (3.49) becomes:

$$v - v_g = \frac{1}{m} \int f \ dt \qquad (3.50)$$

For a translational system, the mass is directly analogous to capacitance. From a circuit viewpoint, the mass always represents capacitance to a fictitious ground terminal. Thus, it is shown dashed in Figure 3.24a.

A rotating mass, Figure 3.24b, is also related to its signal variables by a specialization of Newton's second law. Thus:

$$T = J \frac{d\omega}{dt} \qquad (3.51)$$

where J is the polar moment of inertia $(kg \cdot m^2)$. With a
fictitious non-rotating ground terminal Equation (3.51)
becomes:

$$\omega - \omega_g = \frac{1}{J} \int T \, dt \qquad\qquad (3.52)$$

and the rotating mass also represents capacitance to a
fictitious ground terminal. For the rotational system,
the capacitance equals the polar moment of inertia.
Figure 3.24 shows the circuit representation.

The energy stored by a mass in a translational system
depends on velocity and mass. For a rotational system
the energy stored is a function of angular velocity and
moment of inertia. The formulation for the energy is the
same as that given for the electrical capacitor in Equation
(3.15).

3.5 THERMAL CIRCUIT COMPONENTS

Since the dissipation of energy in all of the systems
considered so far appears as heat, it would be logical to
extend this modelling process to thermal circuits. The
transfer of heat is a complex process which may occur
through a combination of conduction, convection and
radiation. Conduction is heat transfer by direct molecular
communication. In convection, heat is transported by the
movement of fluid masses. Radiation heat transfer, such
as the transfer of heat from the sun to the earth, operates
by a wave motion. We restrict our consideration here to
conduction and convection since these processes fall more
naturally into the linear circuit models (Equation (3.1)).
In this case, we apply Fourier's law of heat conduction
and Newton's law of cooling to obtain the appropriate
network models.

Thermal Signal Variables

We may recognize immediately that the flow of heat, ϕ,
(for example through the walls of a building) is a through
variable. However, it is significantly different than the

other through variables that we have encountered in the
electrical, fluid and mechanical circuits. The difference
is that the heat flow being the rate at which heat energy
is transferred has the units of power by itself. As a
result, the product of the through and across variables is
of no particular significance and the across variable is
chosen without any regard to it. The most convenient
across variable is temperature difference, $\Delta\theta$, expressed
in degrees Celsius.

Despite the differences noted above, the analogy between
the thermal and electric circuit is very real and useful.
The quantity of heat is directly analogous to the quantity
of charge. Thus, the rate at which heat flows corresponds
to the rate at which charge flows, namely the electric
current. An additional similarity is that temperature is
proportional to the energy per unit of heat whereas
voltage is the energy per unit of charge.

Thermal Resistance

Thermal resistance models the relationship between
temperature difference and heat flow rate for a material
of a specific geometry when there is no storage of heat
taking place. The absence of storage is merely an approximation which depends on the properties of the material.
We will discuss this point more thoroughly at the end of
this section. From Fourier's law of heat conduction in
the steady state (no storage) we find that

$$\phi = \frac{kA}{\ell} (\theta_1 - \theta_2) \tag{3.51}$$

where ϕ is the heat flow in watts, θ is the temperature
in °C, k is the thermal conductivity of the material in
W/m·°C, A is the area through which heat flows in m^2,
and ℓ is the length of material, m.

Let us clarify thermal resistance by referring to the
specific geometry shown in Figure 3.25. Here, a round
piece of insulating material has a diameter, d, and
length, ℓ. The piece is surrounded by a collar of perfect

insulation that is shown cross-hatched. The purpose of
the collar is to keep the heat flow in the axial direction.
Though perfect insulation is not physically realizable, it
can be closely approximated. In this case, it is a valuable
concept because it helps to clearly define the area through
which heat flows. The left face of the piece is at temperature, θ_1, the right face at temperature, θ_2, and the ambient
temperature is at θ_a. For this configuration, Equation
(3.51) can be placed in the form of Equation (3.1a) as:

$$\theta_1 - \theta_2 = \left[\frac{4\ell}{\pi k d^2}\right] \phi \qquad (3.52)$$

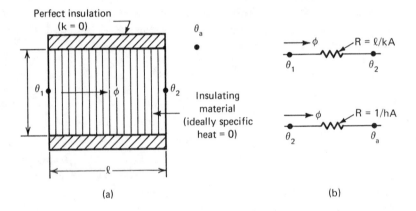

Figure 3.25   Thermal resistance

Thus the thermal resistance of the cylindrical piece with
axial conduction heat transfer is:

$$R = \frac{4\ell}{\pi k d^2}, \quad (°C/W) \qquad (3.53)$$

There is heat transferred from the right face of the
piece to the ambient air at temperature θ_a. This heat
transfer takes place by convection. From Newton's law
of cooling we may express this heat transfer by:

$$\phi = hA(\theta_2 - \theta_a) \qquad (3.54)$$

where h is the surface coefficient (or film conductance) in W/m$^2 \cdot$°C. For the convection heat transfer the thermal resistance has the form:

$$R = \frac{1}{hA} = \frac{4}{\pi hd^2} \ , \ (°C/W) \tag{3.55}$$

The surface coefficient, h, depends on the motion of the fluid medium and is given in heat transfer texts for various geometric configurations and surface conditions. Figure 3.25b shows the resistance circuit for the solid piece and for the film. Together they may be modelled as two resistances in series.

Thermal Inertance

From the signal variables for thermal circuits and Equation (3.1b), we require a relation of the type

$$\theta_1 - \theta_2 = K_2 \frac{d\phi}{dt} \tag{3.56}$$

This equation would define thermal inertance. However, there is no physical situation known to produce such a relation between the temperature difference, $\theta_1 - \theta_2$, and the heat flow rate, ϕ. As a result there is no thermal component analogous to an inductor.

Thermal Capacitance (Heat Storage)

The storage of heat within a mass has long been recognized as a physical process in which the heat energy is regarded almost as a fluid, or as charge as noted previously. As a result, the terms "thermal mass" and "thermal capacity" appear frequently. The basic equation of heat storage in a mass is:

$$\phi = \rho cV \frac{d\theta}{dt} \tag{3.57}$$

where ρ is the mass density, kg/m^3, c is the specific heat of the mass, (W$\cdot$s/kg$\cdot$°C), and V is the volume of the mass, m^3.

The temperature, θ, in Equation (3.57) refers to the temperature of the mass. For pure heat storage components the temperature is uniform throughout the mass. Thus there is only one temperature associated with a mass storage element. To put the mass into the form required for two terminal capacitive components we must assume a fictitious ground terminal at constant temperature, θ_g. Then Equation (3.57) is written as:

$$\theta - \theta_g = \frac{1}{\rho c V} \int \phi \, dt \tag{3.58}$$

and by comparison with Equation (3.1c) we find that the thermal capacitance, C, is equal to:

$$C = \rho c V, \ (W \cdot s / {}^\circ C) \tag{3.59}$$

Consider the cylindrical mass of material enclosed by perfect insulation in Figure 3.26a. For pure heat storage the temperature of the mass is uniform throughout. This means that materials used for heat storage will be very good heat conductors. The thermal conductivity of such materials will be high and they will have very low thermal resistance. If heat is applied to the cylindrical mass the temperature of the mass increases in accordance with Equation (3.58). Figure 3.26b shows the analogous circuit component. Once again, the assumption of a fictitious potential terminal leads to a capacitance to ground and is indicated by a dashed line. All thermal capacitance has this fictitious terminal and therefore, always represents capacitance to ground in the circuit sense. Note that because the terminal is fictitious there is no heat transferred to ground.

If the insulation does not completely surround the heat storage component as shown in Figure 3.27a, the capacitance effect is still to ground. The heat flow that acts to raise the temperature, θ, is now $\phi_1 - \phi_2$ and the component may be represented as shown in Figure 3.27b.

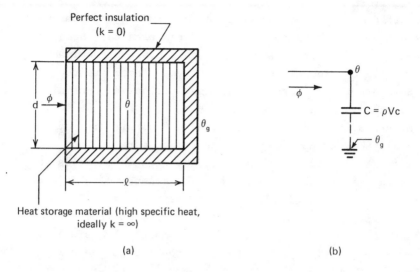

Figure 3.26 Thermal capacitance completely insulated

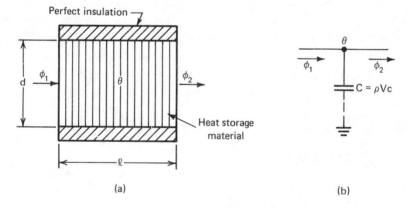

Figure 3.27 Thermal capacitance partially insulated

General Thermal Components

Actually, all thermal components have both resistance and capacitance. The magnitude of the resistance is inversely proportional to the thermal conductivity of the material. The magnitude of the capacitance depends on the density-

specific heat product. The physical size of the component
is a factor in both resistance and capacitance. Some combinations of material and configuration are predominantly
resistive or predominantly capacitive. These are the
models previously described in the ₌ections on Thermal
Resistance and Thermal Capacitance.

Figure 3.28, however, shows the general circuit model
for all thermal components. The temperatures θ_1 and θ_2
are the temperatures of the left and right boundaries and
θ is the mid-plane temperature. The general circuit model
is a TEE-junction with half the resistance on each side
of the mid-plane and a capacitance to ground from the mid-
plane temperature. When the thermal material has a large
thermal conductivity the component reduces to the circuit
shown in Figure 3.27b. In this case, the resistance is
negligibly small and the component is a capacitance to
ground. On the other hand, a material with a lower thermal
conductivity has resistance. If the specific heat-density
product is low the capacitance will be small and the component is purely resistive as shown in Figure 3.25b.

In the general case, however, both the thermal conductivity and the density-specific heat product must be
considered to model a thermal component.

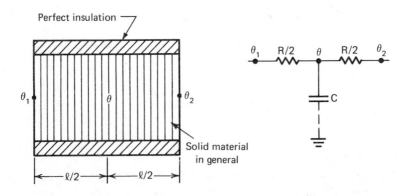

Figure 3.28 General thermal circuit model

3.6 SOURCES AND LOADS

The elemental situations and devices discussed so far,
have implied a source of energy which is quite independent
of the element. These elements have been either dissipative
or storage devices which can only act on the energy while
it is within a system and they are therefore called "passive
elements". The passive elements (resistance, inductance,
and capacitance) are often also called "loads". In con-
trast, "active" elements are devices which imply situations
where energy is supplied to or taken from the system. Often
active elements involve some energy conversion process such
as takes place in a battery, pump, or furnace. That is,
the source of energy for any physical system is modelled
by means of an active element. Thus, in general, a source
supplies energy and a load receives energy. In this section
we will discuss various types of sources and loads.

Across Variable Sources

The details of the origin of the energy are often not
required; only the form in which it is fed into a system
is necessary. For example, most batteries are capable of
supplying energy to an electric circuit with the voltage
difference between its terminals changing only slightly
if the current changes. The ideal situation would pre-
sumably be one in which the voltage difference would be
constant, irrespective of the current flowing from the
battery. In electric circuit analysis, this leads to the
idea of the "ideal" voltage source usually referred to
simply as a "voltage source". Pursuing our across and
through variable analogs, we would expect to find ideal
pressure sources, velocity sources, and temperature sources,
which for convenience can be called "across variable" or
"potential" sources. Figure 3.29a shows the static characteristics (across variable-through variable relation) of
an ideal across variable source. Such a source can supply
flow without changing potential. The circuit representation

of an ideal potential source (Figure 3.29b) is a circle
with polarity indicated by a plus sign.   When the polarity
is reversed the device is sometimes termed a "sink".   However, we may also look on this situation as a negative
source.

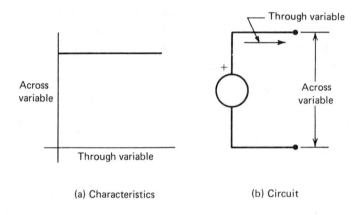

(a) Characteristics   (b) Circuit

Figure 3.29   Ideal across variable (potential) source

Of course, real sources would deviate from the ideal in
that a normal pressure pump, for example, does not quite
maintain constant pressure, irrespective of the flow rate.
However, just as a "real" battery is regarded as a constant
voltage with the "internal resistance" connected in series
(modelled by a voltage source and series connected resistance), the analogous energy sources can be modelled by
an across variable source and series dissipative element.
Figure 3.30a shows the static characteristics of a real
across-variable source.   The equivalent circuit representation of the real source is shown in Figure 3.30b.   The
circuit representation has an ideal source and an internal
resistance, R_i, in series.
It should be noted that for dynamic studies the model of
an energy source may require at least one storage element
in addition to the across variable source and dissipative

element. This should become evident when Thevenin's and Norton's network theorems are considered in Chapter VII.

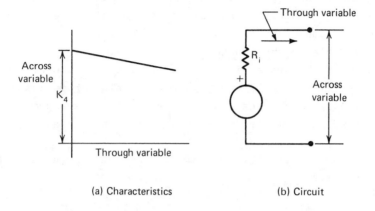

(a) Characteristics   (b) Circuit

Figure 3.30 Real across variable (potential) source

Situations involving an ideal velocity source are not often encountered in elementary mechanics. The main point is to look for whether the velocity at a point in the system is constrained to some specific value or function of time. The situation where a body is struck and carried by a much larger mass which is already moving at a constant velocity can be modelled by a velocity source since the velocity of the larger mass is virtually unchanged after making contact with the smaller mass. The resulting forces may be sufficiently large to cause damage, and therefore great care is required when subjecting any part of a mechanical system to a suddenly applied constraint of constant velocity.

The ideal temperature source is a reasonable model of a thermostatically controlled oven, provided the item to be heated can be inserted without any change in oven temperature. This is inevitably very difficult to achieve exactly. Nevertheless, the times involved are such that in many such situations the model of the temperature-

controlled oven as an ideal temperature source gives
reasonably accurate results.

In general, an ideal across variable or potential source
is the correct model for any situation in which the across
variable is constrained to some specified value or function
of time. This is true whether the constraint is produced
naturally or whether it is the result of some automatic
control scheme. If the end result is a value or functional
expression for the across variable which is independent
of the through variable, it is appropriate to use an ideal
potential source as model.

The characteristic equation of a potential source may
be written as:

$$(\text{Across Variable}) = K_4 - R_i(\text{Through Variable}) \qquad (3.60)$$

where K_4 is the value of the across variable when there is
no through variable. Note that for the ideal potential
source, $R_i = 0$, and the across variable is constant and
independent of the through variable.

Through Variable Sources

As might be expected, situations arise in which the value
of a through variable is constant or is some specified
function, independent of across variable. Most elementary
mechanics are based on the premise that it is possible (at
least theoretically) to apply a force which is independent
of the velocity of the point at which it is applied. Such
situations can be modelled as an ideal force source usually
called simply a "force source". In electrical engineering
there are very few situations where the current from a
source is naturally constant, but it is quite common for
the current from a DC power supply to be controlled elec-
tronically such that these supplies may be represented by
current sources. The ideal current source is one where
the current is entirely independent of the voltage across
it. Figure 3.31a shows the static characteristics of an

ideal through variable or flow source. The through variable remains constant for all values of the across variable. The circuit symbol to represent the ideal flow
source is a round circle with an arrow indicating the
flow direction along the diameter.

At this stage, it may be wise to emphasize the fact that
a "source of current" and a "current source" have sig-

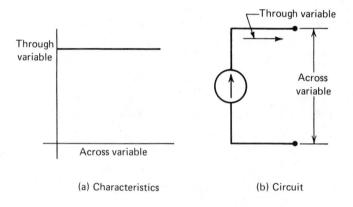

(a) Characteristics   (b) Circuit

Figure 3.31 Ideal through variable (flow) source

nificantly different meanings. A "source of current" has
absolutely no implication as to whether the voltage or
current or neither is constant. However, "a current source"
unequivocally implies an ideal flow source, namely one
in which the current is constant or some function independent of the voltage. Similarly, a "source of force"
must not be confused with a "force source".

For convenience, sources in which the through variable
is constrained are called "through variable sources" or
"flow sources". This latter term is particularly applicable to hydraulic pumps in which clearly identifiable
quantities of liquid are passed through the pump. Provided
such pumps are driven at constant speed, the rate at which
the liquid is delivered is constant. In the case of thermal

systems, an ordinary heater without any thermostatic control
produces heat at a rate which is reasonably independent of
the temperature. An ideal flow source is therefore used
to model this situation.

In general, a flow source is used to model any situation
in which the through variable is constrained to some specified value or function of time. This is true whether
the constraint is produced naturally or whether it is the
result of some automatic control scheme. If the end result
is a value or functional expression for the through variable which is independent of the across variable, it is
appropriate to use an ideal flow source as a model for
this situation.

Actually, there are no ideal flow sources. Any real flow
source is affected by the across variable. Thus for
example, a constant displacement pump would slow down if
the pressure difference got too large. Figure 3.32a shows
the static characteristics of a real flow source. As a
circuit component the real flow source may be considered
an ideal flow source in parallel with an internal resistance, R_i. This is shown in Figure 3.32b. The equation
of the real flow source is:

$$\text{Through Variable} = K_5 - \frac{1}{R_i} [\text{Across Variable}] \qquad (3.61)$$

where K_5 is the value of the through variable when the
across variable is zero. In this case, an ideal flow
source has an infinite value of internal resistance. Note
that under this condition the through variable does not
depend on the across variable.

The characteristics of real flow sources and real potential sources are very similar. (Compare Figures 3.30a
and 3.32a). For a unity value of the internal resistance
Equations (3.60) and (3.61) are in the same form and can
be identical when $K_4 = K_5$. In this case, we would not
designate the component as a "flow" source or a "potential"
source. We would merely designate it as a "source".

In general, any active element can be used to model
either a source or a sink of energy. If the device is
acting as a source, positive flow will emerge from the
positive terminal for both the potential and flow sources.
If the physical situation is one which absorbs energy,
positive flow will enter at the positive terminal. Although
not entirely consistent, it is nevertheless, convenient to
call the active element a source, whether it actually is
acting as a source of energy or whether it is acting as a
sink.

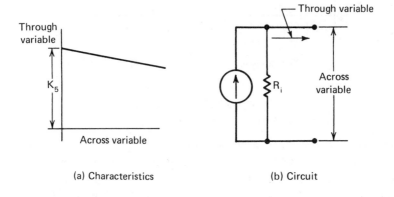

(a) Characteristics   (b) Circuit

Figure 3.32 Real through variable (flow) source

Loads

Resistance, inductance, and capacitance are all load
elements. However, resistance is the only load that has
a static characteristic. Inductance and capacitance
generally have time dependent signal variables. Figure
3.33a shows the characteristics of a resistive load. For
a linear resistance the characteristic is a straight line
passing through the origin. The slope of the line is the
value of resistance. Figure 3.33b indicates the analogous
circuit for the load resistance, R_L. The relation between

the variables for a resistive load can be obtained from
Equation (3.1a) for this case as:

Across Variable = R_L (Through Variable)   (3.62)

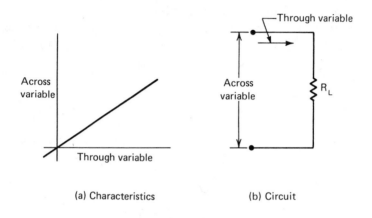

(a) Characteristics   (b) Circuit

Figure 3.33   Resistive load

The interaction of a source and a load involves some
principles of circuit theory. Although circuit theory
is not considered until Chapter IV, we may gain some
insight now by superimposing source and load characteristics. For example, suppose an ideal potential source
and resistive load characteristics are superimposed
(Figure 3.34a). This is equivalent to the circuit connecting an ideal source to a resistive load (Figure 3.34b).
Let us recall that we may interpret a source as representing the locus of all possible operating states of the
energy element. Similarly, the load represents the locus
of all possible values of the signal variables for a
resistance. The circuit configuration is such that both
loci must be applicable simultaneously. The only point
that permits both loci to co-exist is the intersecting
point of the characteristics. This point is called the
operating point because it shows the value of the signal

variables in the circuit. This is a graphical interpretation of the solution of two simultaneous equations
(i.e. Equations (3.60) and (3.62)).

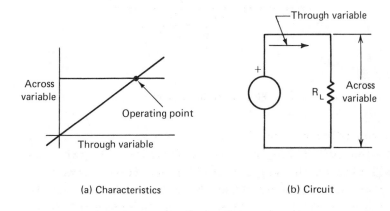

(a) Characteristics   (b) Circuit

Figure 3.34 Source-load interaction

The source-load method is especially valuable in non-
linear circuits where the results are more difficult to
obtain analytically.

Modelling of Coulomb Friction

When considering friction, it was noted that the characteristic of Coulomb friction is that the frictional force
is independent of velocity. Accordingly, the appropriate
model for this situation is a force source (torque source
if rotating) connected such that the flow of force enters
at its positive terminal. This model is exact provided
motion or rotation is restricted to one direction only.
If the motion reverses, the polarity of the force source
must be reversed since the frictional force reverses.

Let us examine, for example, the case of a mass, m,
moving along a rough horizontal surface with a velocity,
v (Figure 3.35). The normal force between the mass and
surface, N, is equal to the weight, mg. The frictional

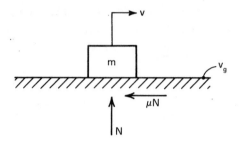

Figure 3.35 Sliding friction

force retarding motion, f, is μN (or μmg) where μ is the
coefficient of sliding or kinetic friction. Figure 3.36
shows a plot of the velocity difference-force characteristics of the friction. When the velocity difference is
zero (no motion) the frictional force is somewhat larger
than when the mass is moving. This occurs because the
coefficient of static friction is larger than the coefficient of kinetic friction. Actually, the frictional
force decreases slightly with increased velocity difference. Notice, however, that the characteristic shown in
Figure 3.36 is quite similar to the ideal flow source
characteristic shown in Figure 3.31a. As an approximation,
therefore, we use a force source circuit model to represent
Coulomb friction. If the motion reverses in direction,
the force source is reversed. This is a non-linear effect
which is outside the scope of this text, the analysis of
such a situation usually being introduced in texts on
automatic control systems.

Modelling of Weight

Another situation involving a constant force which is
not necessarily a source of energy is the weight of a
mass subject to vertical motion. In Figure 3.37a, an
applied force, f, is required to balance the weight if
the mass is to remain stationary, this being the means
by which we are aware of the gravitational force on an

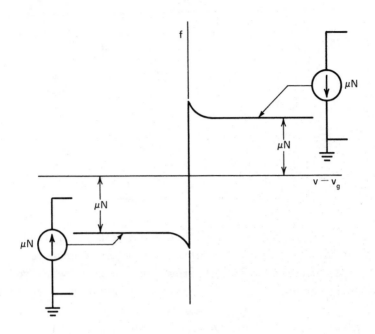

Figure 3.36 Modelling of Coulomb friction by force (flow) source

object. However, the weight of the object is the gravitational force, mg, which should be expressed in newtons.
Provided distances from the centre of mass of the earth
do not vary significantly, the weight is constant and can
therefore be modelled by means of a force source. The
polarity of the source will depend on the direction chosen
for positive velocity. If velocity is positive when the
motion is upwards, the polarity is that shown in Figure
3.37b.

3.7 ELEMENTS AS MATHEMATICAL OPERATORS

From the elemental equations of all of the passive elements, it can now be noted that the through variable is
either a multiple, or time-integral or time-derivative of
the across variable. That is, the voltage drop across a

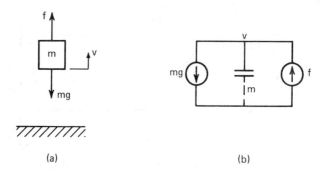

Figure 3.37   Modelling of weight

resistor, for example, is simply the current multiplied
by a constant (i.e., the resistance). This is true for
ideal resistance no matter how the current is varying and
therefore the appearance of a graph of voltage drop against
time is identically the same as that of a graph of current
against time, other than the scale factor which the value
of resistance introduces. The voltage difference across
an inductor is the time derivative of the current flowing
through it with the scale factor of the value of inductance included. If the current is constant, the voltage
difference across an ideal inductance is zero. If the
current is some function of time, the voltage difference
is then the derivative of that function multiplied by the
value of inductance as a scale factor. The voltage difference across a capacitor on the other hand, is the time
integral of the current flowing through it with the scale
factor of the capacitance included. If the current is
constant, the voltage drop across the capacitor will
increase at a uniform rate. If the current is some function of time, the voltage drop is then the integral of
that function divided by the capacitance.

The process of integration generally requires evaluation
of the constant of integration. However, in this context,
the value of the integral is usually known at the instant
in time at which the process is considered to start, i.e.,

t = 0 in most cases. By using the format of a particular
integral the value of the constant of integration usually
is obtained automatically. For example, if we consider a
capacitor which is initially charged when a current of
$Iu_S(t)$ is passed through it, the resulting voltage drop
across the capacitor may be developed from the basic
Equation (3.1c):

$$\Delta e(t) = \frac{1}{C} \int_{-\infty}^{t} i \ dt$$

$$= \frac{1}{C} \int_{-\infty}^{o} i \ dt + \frac{1}{C} \int_{o}^{t} i \ dt \qquad (3.63)$$

The first term of the right-hand side represents the
voltage drop (Δe_0) resulting from the initial charge on
the capacitor. Thus:

$$\Delta e(t) = \Delta e_0 + \frac{1}{C} \int_{o}^{t} i \ dt$$

$$= \Delta e_0 + It/C \qquad (3.64)$$

Circuit models for initial conditions will be presented in
Chapter IV. However, we should be aware that time integration always adds the integral in an interval to the
value of the variable at the beginning of the interval.

Example 3.1

If the effective driving force on an automobile has
the form shown in Figure 3.38a, determine the resulting
velocity if the road is horizontal and wind resistance
is neglected, and the mass of the automobile is 1500 kg.
The automobile is initially at rest.

Solution

The graph of force against time has three basic segments where it has a value other than zero. The first
segment, from t = 0 to t = 5 has a constant slope 240 N/s,

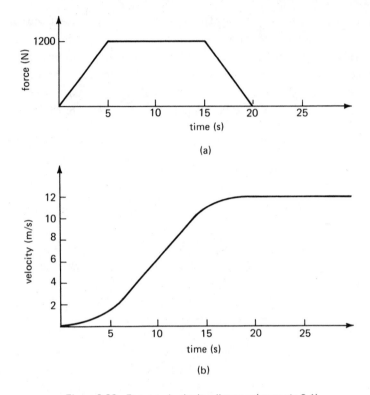

Figure 3.38   Force and velocity diagrams (example 3.1)

the second from t = 5 to t = 15, has a constant value
of 1200 N, and the third, from t = 15 to t = 20, has a
constant slope of -240 N/s.

The elemental equation describing the situation is:

$$f = m \frac{dv}{dt} \qquad \text{or} \qquad v = \frac{1}{m} \int f \, dt$$

Thus the velocity is the integral of the function shown
in Figure 3.38a divided by the scale factor 1500. This
is a case where the form of the velocity is obtained quite
simply by noting that the integral of a straight line of
non-zero slope is a parabola whose change in ordinate
over the period under consideration equals the area under
the existing function. Thus the velocity at t = 5 s is:

$$v(5) = 0 + \frac{1}{1500} \, (1/2)(5)(1200) = 2 \text{ m/s}$$

since the vehicle was initially at rest.

The velocity during the second segment is a straight line of slope $1200/1500$ or 0.8 m/s^2, the value at the end of the period being obtained from the increase in the area underneath the graph of the force.

$$v(15) = 2 + \frac{1}{1500} \, (1200)(10)$$
$$= 2 + 8 = 10 \text{ m/s}$$

Similarly, the velocity during the third segment is a parabola, concave down, the final value being

$$v(20) = 10 + \frac{1}{1500} \, (1/2)(5)(1200)$$
$$= 12 \text{ m/s or } 43.2 \text{ km/h}$$

After t = 20 s, there being no further change in the area under the graph of force, there is no further change in velocity, which, with the assumption of no wind resistance, would remain constant. Hence, the sketch of velocity is that shown in Figure 3.38b.

Example 3.2

A frictionless pipe of 1.0 m diameter carries water from a reservoir to a turbine over a distance of 10 km. The flow rate is 20 m^3/s and the flow is to be cut off in a period of 10 seconds, as shown in Figure 3.39a. Determine the pressure difference due to the inertia of the water.

The fluid inertance is given by:

$$L = \frac{\rho \ell}{A}$$
$$= \frac{(1000)(10)(10)^3}{(\pi)(1^2)/4}$$
$$= 1.27 \, (10)^7 \, (\text{N} \cdot \text{s}^2/\text{m}^5)$$

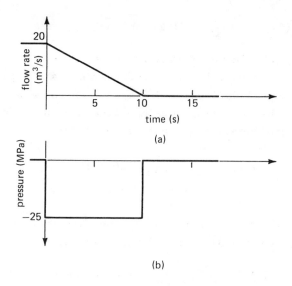

Figure 3.39 Flow rate and pressure drop across frictionless fluid line (example 3.2)

while the flow is being cut off, the rate of change is:

$$\frac{dq}{dt} = -\frac{20}{10} = -2 \text{ m}^3/\text{s}^2$$

The pressure difference due to the inertia effect is therefore:

$$\Delta p = L \frac{dq}{dt}$$

$$= (1.27)(10)^7(-2)$$

$$= -25.4(10)^6 \text{ Pa}$$

The pressure drop is shown in Figure 3.39b.

It should be noted that in effect, this is the pressure acting on the gate or valve as it is being closed. Just before final closure, it is therefore acting over almost the entire cross-sectional area of the pipe, thus applying a force of

$$f = (\Delta p)A = \frac{(25.4)(10)^6 (\pi)(1)^2}{4}$$

$$= 20(10)^6, \text{ N}$$

which is the same as the gravitational force on a mass of $2.04(10)^6$ kg. This extremely high force indicates that damage to valves may result from rapid closure. Indeed, unless special precautions are taken the closing of a valve in a pipe which has a moving liquid generally results in forces which may be large enough to displace the pipe with the possibility of resulting damage. This phenomenon is usually called water hammer.

3.8 SUMMARY

The process of determining analogous physical situations can now be seen to be in two main parts. The first has been to identify the analogous through and across variables. Although at first, selected on the basis that the product would be power, there is no reason not to use power as the through variable, if the elemental equations turn out to have identically the same form, as in the thermal system. The final choice is shown in Table 3.1.

The elemental physical situations are modelled by means of the circuit components. With the systems having power as the product of the through and across variables (electric, fluid and mechanical), there are only three components. One dissipates energy, usually as heat, and the two others store energy. One of the storage elements stores energy when the across variable exists, whether or not the through variable has some value. For convenience, these components can be categorized as A-type storage elements. The other element stores energy by virtue of the through variable whether or not the across variable has some value. For convenience, these components can be categorized as T-type storage elements.

TABLE 3.1   THE ANALOGOUS THROUGH AND ACROSS VARIABLES

	Through	Across	Product
Electric Circuit	Current (A)	Voltage (V)	Power (W)
Fluid	Volume Flow Rate (m^3/s)	Pressure (Pa)	Power (W)
Mechanical Rotation	Torque ($N \cdot m$)	Angular Velocity (rad/s)	Power (W)
Mechanical Translation	Force (N)	Velocity (m/s)	Power (W)
Thermal	Heat Flow Rate (W)	Temperature Difference (°C)	---

For thermal systems, there are only two basic elements:
(a) dissipative, and (b) A-type storage.
The basic forms of elemental equations are:

(Across Variable) = K_1 (Through Variable)

(Across Variable) = $K_2 \dfrac{d}{dt}$ (Through Variable)

(Across Variable) = $K_3 \int$ (Through Variable) dt

The appropriate interpretation of K_1, K_2, and K_3 is summarized in Table 3.2.

TABLE 3.2 THE ANALOGOUS ELEMENTS

	Dissipative K_1	T-Type K_2	A-Type K_3
Electric	R	L	1/C
Fluid	R	L	1/C
Mechanical (Rotation)	1/B	1/K	1/J
Mechanical (Translation)	1/b	1/k	1/m
Thermal	R	---	1/C

Problems

3.1 The graphs shown in Figure P3.1 represent the through and across variables for particular components. Identify the components and determine their magnitude.

3.2 Water flows into an open tank at $q_1(t)$ and out of the tank at $q_2(t)$ as shown in Figure P3.2. The initial height of the water in the tank is 0.25 m and the cross-sectional area of the tank is 49 cm^2. Sketch a graph representing the pressure, p(t), at the bottom of the tank.

3.3 Some data is presented below from four elemental physical situations. In each case, identify the element and state whether the analogous electric circuit element is resistance, inductance or capacitance. Give the magnitude of the component.

a) Element A

Torque, N·cm	Angular Velocity, rad/s
60	1
120	2
180	3

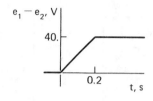

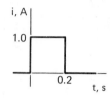

(a)

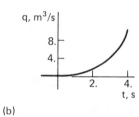

(b)

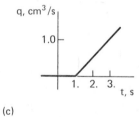

(c)

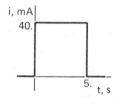

(d)

Figure P 3.1

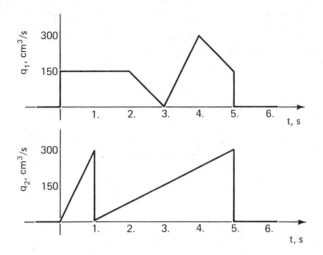

Figure P 3.2

b) Element B

Force, N	Deflection, cm
10	1.25
20	2.50
50	6.25

c) Element C

Pressure Difference, N/cm^2	Volume, cm^3
4	160
12	480
16	640

d) Element D

Force, N	Time, s	Velocity, m/s
9	0	0
9	1	4
9	2	8
9	3	12

3.4 A 90 kg mass rests 165 m from a wall. There is no friction between the mass and the ground. A series of impulses in force are applied to move the mass

toward the wall. The impulses are shown in Figure
P3.4. Find the time for the mass to reach the wall
and its velocity at this time.

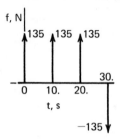

Figure P 3.4

3.5 A mechanical system consists of a mass and a spring
as shown in Figure P3.5. The velocity at the left-
end of the spring $v = 2.5u_s(t)$, m/s. The difference in velocities at each end of the spring is

$$v_1 - v_2 = (2.5 \cos 4t) \, u_s(t), \text{ m/s}$$

The spring constant $k = 80$ N/m. Find:
a) the force in the spring;
b) the value of the mass, m.

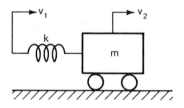

Figure P 3.5

3.6 A flywheel with moment of inertia, J, turns through
an angle $\theta(t)$ where

$$\theta(t) = 8t^2\ u_s(t) - (12t^2 - 120t + 300)\ u_s(t - 5)$$
$$+ (7t^2 - 140t + 700)\ u_s(t - 10),\ \text{rad}$$

When the angular velocity of the flywheel is 200 rad/s the torque required is 72 N·m. Find the value of J.

3.7 The voltage drop, $e_1 - e_2$, across a 2 henry inductor is shown in Figure P3.7. Find the value of the current when t = 3.36 s. What is the energy stored at this time?

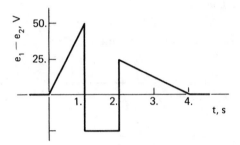

Figure P 3.7

3.8 An angular velocity input is applied to the free end of an angular spring. The other end is firmly attached to a wall. The spring constant k = 8 N·m and the input function is shown in Figure P3.8. Find the magnitude of the torque developed in the spring at t = 12 s.

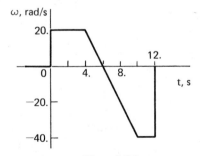

Figure P 3.8

3.9   The current passing through an inductance is sinωt,
       A.   The value of inductance is 0.1 H.   Plot a
       graph of voltage amplitude across the inductance
       against the angular frequency, ω, which varies
       from 0 to 20 rad/s.

3.10 A perfectly insulated tank is filled with 0.05 m³
      of water at 20°C.   An immersion heater in the water
      is suddenly turned on and the heat input to the
      water is shown in Figure P3.10.   If the specific
      heat of water is 4180 joules/(°C·kg) find the temper-
      ature in the tank as a function of time.   To prevent
      the water from exceeding a temperature of 50°C what
      is the longest time the heater may remain on?

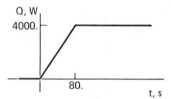

Figure P 3.10

3.11 The voltage drop across a 1 μF capacitor is shown
      in Figure P3.11.   Plot the current in the capacitor
      against time and find the energy stored when t = 5 s.

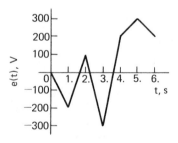

Figure P 3.11

3.12 A mass, m, is falling freely.   When it is a distance,
      x, above ground its velocity is v_0.   Prove that

$$x = v_o t_1 + 1/2 g t_1^2$$

where t_1 is the time for the mass to reach the ground. Assume that air resistance is negligible.

3.13 Measurements on a circular fluid line with water flowing show the pressure drop and volume flow functions plotted in Figure P3.13. The line may be modelled as an inertance and resistance in series.

a) Find the length and diameter of the line.

b) When does the Reynolds number exceed 2 000?

c) How much energy is stored in the inertance when laminar flow ceases?

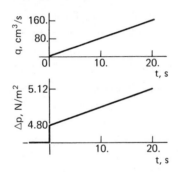

Figure P 3.13

3.14 The graphs shown in Figure P3.14 represent the through variable in a mechanical component. Suppose that the following mechanical components are available.

(1) spring, $k = 80$ N/m

(2) mass, $m = 150$ kg

(3) damper, $b = 40$ N·s/m

Sketch the resulting across variable for each component subject to a), b), and c). Find the value of the across variable at $t = 8$ s.

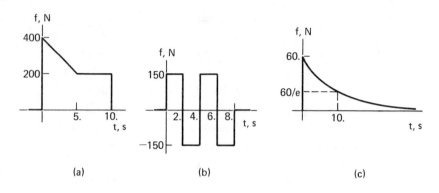

Figure P 3.14

3.15 An open water tank has a square cross-section 7 m
on each side.   The tank is initially empty when the
flow input shown in Figure P3.15 is applied.   The
initial value of the flow input is q_o.   If the tank
is half-full at t = 14.7 s and completely full at
t = 27 s, find the value of q_o and the height of the
tank.   Make a sketch of the pressure at the bottom
of the tank in the time interval from 0 to 30 s.

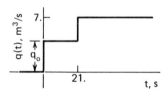

Figure P 3.15

3.16 A flywheel has a polar moment of inertia of 70 kg·m².
A torque is applied to the flywheel as shown in
Figure P3.16.   If the diameter of the flywheel is
1 m, find the total distance travelled by a point
on the outer edge in the time interval from 0 to
60 s.   How much energy is stored in the flywheel
at t = 36 s?

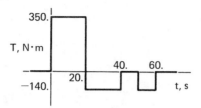

Figure P 3.16

3.17 A car battery with clean terminals can be considered as an ideal voltage source delivering 12 volts. The battery is connected to the ignition circuit of the car through the ignition switch. Assume that the ignition circuit is a pure resistance, R = 3Ω. If the battery terminals become corroded an additional resistance of 1Ω may be considered as the internal resistance of the source. What is the percentage decrease in ignition circuit current due to corrosion?

3.18 A 20 kg mass is moving at a constant velocity of 40 m/s. If the energy stored in the mass is all transferred to a spring with k = 20 N/m, what force would be developed in the spring?

CHAPTER 4

BASIC
NETWORK
MODELS

There are two basic ways to represent an electric circuit
schematically. In one, the details of the connections are
shown. Diagrams of this type are difficult to follow. In
the other, the details of connection are ignored. Instead
the components are arranged for maximum clarity to show
how currents and voltages combine or separate. Schematic
diagrams of this type are the ones normally used for analyzing the performance of an electric circuit. There are
standard techniques for writing equilibrium equations by
direct reference to these diagrams. This enables the
formulation of differential equations to represent the
electrical signal variables.

The interactions of the elemental physical components in
the mechanical, fluid and thermal media may also be

represented by network diagrams having exactly the same
form as the electrical network diagrams. This serves to
make the same analytical techniques available to these
media. The result is that equilibrium equations for systems of mechanical, fluid, and thermal components can be
written with equal facility and reliability.

4.1 FORMULATION OF CIRCUITS

  The objective of circuit formulation is to place mechanical, fluid, and thermal component arrangements into circuit
diagrams. The essence of the problem is to identify those
points in the system that may have different values of the
across variable. In addition to these points each circuit
must have a datum or reference value for the across variable. The datum is usually called the "ground" of the circuit. The other network points are called "nodes". For
mechanical systems the number of nodes (other than the
datum) is equal to the number of "degrees of freedom".
  The process of circuit formulation requires the identi-
fication of the basic elemental effects. Specifically, we
must first find all the across variable nodes and then determine the appropriate element or elements that connect
them together. The nodes therefore become the junctions
of the circuit branches. Each branch indicates one element
and corresponds to one elemental physical situation. The
process is illustrated in the following examples.

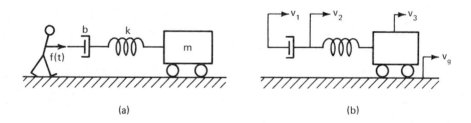

(a)   (b)

Figure 4.1 Mechanical arrangement (example 4.1)

Example 4.1

Consider the mass-spring-damper system shown in Figure 4.1a in which a man applies a force, f(t), to one end of the damper. The first step is to identify all parts of the system which are constrained to move at the same velocity and to indicate the direction in which these velocities will be considered as positive. This step is shown in Figure 4.1b. All the velocities are taken positive in the same direction. This facilitates the formulation of equilibrium equations and makes the results significantly easier to interpret. For the circuit under consideration there are three velocities that are not necessarily the same. These are the velocity at the force end of the damper (v_1), the velocity at the point between damper and spring (v_2) and the velocity of the mass (v_3). Therefore the equivalent circuit diagram will have three nodes or junctions in addition to the one representing the datum (v_g). These junctions, shown in Figure 4.2a, are the framework on which the elements are inserted. The junctions correspond to the variables v_1, v_2, v_3 and v_g. For convenience and clarity the ground terminal, v_g may be extended along a line at the same potential.

The passive components are placed in the circuit diagram by noting the velocity difference (across variable)

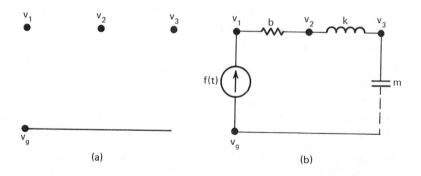

(a)   (b)

Figure 4.2   Mechanical circuit (example 4.1)

that is required for each elemental equation. For example, the damping, b, models an effect which is dependent
on the difference between v_1 and v_2.

Thus the damping element is connected between junctions
1 and 2. Damping, as discussed in Chapter III, is mechanical resistance and is a dissipative component. We represent it schematically by a resistance symbol. The
spring stiffness, k, also models an effect that depends
on velocity difference. In this case, the velocities at
the ends of the spring are v_2 and v_3. The spring element
is therefore connected between these two nodes. Recall,
Chapter III, that the spring is analogous to the electrical inductance and has the same circuit symbol. Now
the physical equation that describes the motion of the
mass (Equation (3.48)) shows that it is dependent on only
one velocity. In this case, the velocity is v_3. Since
all circuit components require two terminals we add a
terminal that does not change the physical relations.
This terminal is the ground junction at zero velocity.
When a fictitious terminal is used the circuit diagram
contains a dashed line to show this. The mechanical mass
is analogous to an electrical capacitance with one terminal always connected to ground. Figure 4.2b shows the
circuit diagram for the mass-spring-damper system.

The driving force exerted by the man is modelled by
a flow source connected between junctions v_1 and v_g.
Since the force is applied in the direction of positive
velocity the flow must be shown as entering junction v_1.
Note that the force representation is complete because it
also indicates the necessary reaction force on the datum
in the opposite direction (away from junction v_g).

The electrical analog of this particular mechanical circuit is a series RLC circuit with a current source input.
However, it must be emphasized that other mass-spring-
damper systems in which the elements appear in tandem do
not necessarily result in a series circuit. This point is
illustrated in Example 4.2.

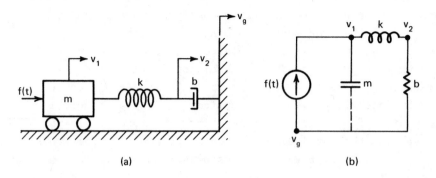

Figure 4.3 Mechanical circuit (example 4.2)

Example 4.2

Figure 4.3a shows a mass-spring-damper arrangement in which the force is applied to the mass. Here again, the mechanical arrangement has the components in tandem. However, in this case, the order is different. With this arrangement there are now only two velocities that can be identified as independent. These are the velocity of the mass (designated as v_1) and the velocity of the point between the spring and damper (designated as v_2). Once again, the reference terminal is the velocity of the ground (v_g). The mass represents a capacitance between v_1 and v_g. The spring is equivalent to an inductance between v_1 and v_2, and the damper is a resistance that goes from v_2 to v_g. The resulting circuit is shown in Figure 4.3b.

In this case, the passive elements do not form a series circuit. Thus, it is important not to jump to conclusions about series or parallel equivalences solely on the basis of mass-spring-damper sketches. The analogous circuit representation can be obtained only by determining the terminals and connecting them with the appropriate component.

Before proceeding to more complicated examples with more than one source, it will be useful to discuss further the

polarity of sources. Polarity concepts are especially
pertinent when the sources are not all acting in the same
direction. Consider the isolated mass shown in Figure 4.4,
in which several possible situations are indicated. In all
cases, the direction selected for velocity (v) is concep-
tually positive. Then the application of the elemental
equation (f = m dv/dt) requires that the force, f is positive from the terminal v through the element to the ground
terminal. Thus, the situations presented in Figure 4.4a and
b, are identical in the sense that the force applied to the
mass acts in the same direction as the assumed positive
direction for velocity. In other words, the force tends
to increase the velocity. On the other hand, the force in
Figure 4.4c is directed to the left while the assumed positive velocity direction is to the right. As a result, the
corresponding circuit representation shows a force source
that is directed away from the terminal.

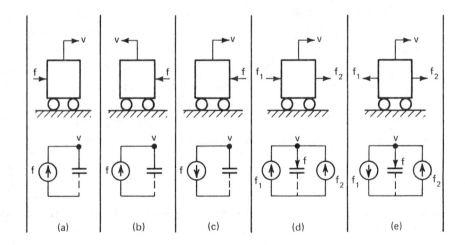

Figure 4.4 Polarity of sources

When two forces act on a mass the circuit has two sources
and therefore contains three branches which meet at a
junction. In the case of two forces in the same direction

(Figure 4.4d), the previous discussion indicates that each source should have its positive flow toward the junction. The reason is, of course, that both forces act to increase the velocity. Continuity of flow requires that the total force passing through the mass is the sum of the flows entering the junction. That is, $f = f_1 + f_2$. In conventional mechanics terms the sum of the unbalanced forces determines the acceleration of the mass. The circuit approach therefore is merely another way of recognizing and developing basic physical relations.

Suppose now that the forces are acting in opposite directions, as in Figure 4.4e. The force source in the direction of positive velocity (f_2), is modelled by a flow source directed toward the terminal v. This is the force that tends to increase velocity. The opposite force, f_1, is modelled by a flow source directed away from the terminal and has a flow at the terminal that yields $f = f_2 - f_1$.

Both situations (Figures 4.4d and e) are in exact agreement with physical reality. The junction of the circuit representation corresponds to the free body diagram of mechanics. However, for complicated mechanical arrangements, several free body diagrams may be required. An advantage of the network representation is that all the junctions are interconnected and the interactions between different parts of a mechanical arrangement are completely described by only one diagram.

For situations which are modelled by velocity sources the procedure is simply to observe whether the source is in the positive or negative direction. If positive, the positive terminal of the velocity source is connected to the junction. When a velocity source is directed opposite to the assumed positive velocity direction the negative terminal of the source is connected to the junction. These points are illustrated in the following examples.

Example 4.3

Figure 4.5a shows an automobile approaching a bump in the road. The suspension system of the automobile sup-

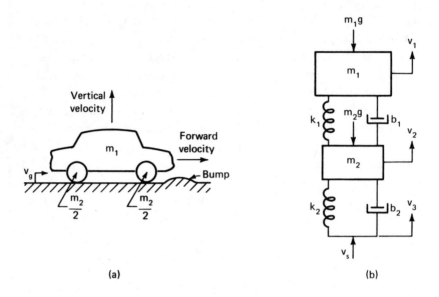

(a)   (b)

Figure 4.5 Mechanical arrangement (example 4.3)

ports the car body by means of a set of springs (k_1) and
shock absorbers (b_1). The tires exhibit a spring effect
(k_2) which is close to the ideal spring but the hysteresis
normally associated with rubber springs is modelled by an
equivalent damper (b_2). The constraint at the point of
contact between the tire and the road is that the position of this point is defined. If the car is driven at
constant forward velocity along the road, the vertical
velocity of the point of contact is defined. This vertical
velocity constraint is, therefore, appropriately modelled
by a velocity source (v_S). In addition, there are force
sources to model the weight of the car and the weight of
the wheels. These considerations lead to the mass-spring-
damper arrangement shown in Figure 4.5b.

To obtain the network diagram we must first determine
the number of independent velocities associated with the
mechanical sketch. Each mass may move independently and
this leads to terminals representing v_1 and v_2. The
point of contact with the ground is v_3. The velocity

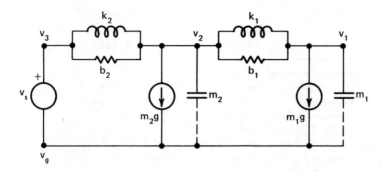

Figure 4.6 Mechanical circuit (example 4.3)

of the ground is v_g. Thus, there are three terminals
plus a ground on the circuit diagram (Figure 4.6). The
connection of components between the terminals is obtained
by referring to the sketch in Figure 4.5b. The velocity
source, v_s, acts in the direction of positive velocity.
As a result the positive terminal of this source is
connected to v_3. This means that $v_s = v_3 - v_g$. Had
the reference velocity direction been selected as down
the negative terminal of v_s would have been attached to
v_3. In that case, $v_s = v_g - v_3$. The force sources due
to the weight of the car ($m_1 g$) and the weight of the
wheels ($m_2 g$) are directed away from their respective
terminals (v_1 and v_2). Obviously, if the reference
velocity was down these sources would be pointed toward
the terminals.

Example 4.4

Winds apply large forces to building structures. These
forces must be included in the structural design. If the
wind is to be treated as a variable force source, the
dynamic problem posed is rather difficult. However, a
lumped model of the structure can simplify the problem
appreciably. Figure 4.7a shows a schematic drawing of
a three floor building subjected to a wind force. We
wish to obtain a circuit diagram for this situation.

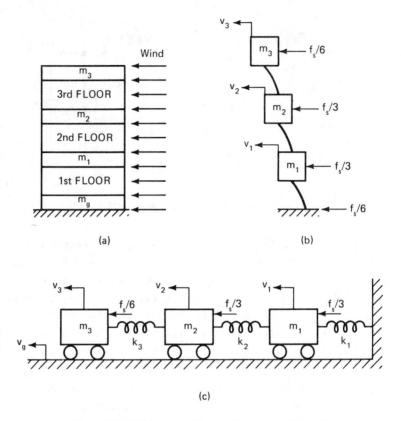

Figure 4.7 Building model for wind gusts (example 4.4)

In this example, the first step is to develop a mechanical sketch. To do this, we must have some idea
about the distribution of mass in the structure. We
will assume here that most of the mass of the building
is concentrated in the slabs between floors and ceilings.
Thus m_3 lumps the mass of the roof with a portion of the
mass of the supporting columns. Similarly, m_2 lumps the
slab, between the ceiling of the second floor and the
floor of the third floor, with a portion of the column.
The mass, m_1, is also lumped in a similar manner. Now
the slab on which the building is constructed may be
modelled as m_g. However, this lump is fully constrained
by the ground. Thus, we may model the structure by three

lumped masses m_1, m_2 and m_3. The columns connecting these masses are considered as cantilever beams which have an equivalent (spring) stiffness.

Figure 4.7b is an exaggerated sketch of the building subjected to a total wind force of f_s. The wind force is assumed to be uniformly distributed across the entire outside wall. In relation to the concentration of mass, however, we assume that m_1 and m_2 are each acted upon by $f_s/3$. The remaining $f_s/3$ is assumed to be equally distributed between m_3 and m_g. Thus, m_3 is acted upon by $f_s/6$.

Figure 4.7c emphasizes the fact that the motion under consideration is entirely transverse motion. The cantilever springs are modelled by ordinary extension springs of the same stiffness.

Once the sketch shown in Figure 4.7c is made, we may proceed as before to identify the independent velocities. There is always a velocity associated with each mass so we have three terminals plus ground to form our circuit. Figure 4.8 shows the circuit model for the building subjected to wind. Note that the masses are always connected to ground. The wind forces act in the assumed positive velocity direction and therefore the arrows in the flow sources point toward the terminals.

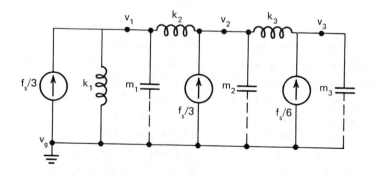

Figure 4.8 Mechanical circuit (example 4.4)

4.2 BASIC CIRCUIT LAWS

The objective of circuit analysis is to derive mathematical relations between output variables and input variables. The circuit representation of a physical system
permits us to accomplish this analysis through the application of two basic circuit laws. One of these laws,
Kirchhoff's current law (KCL), refers to conditions at
any junction and is expressed as:

$$\Sigma(\text{Through Variables}) = 0 \qquad (4.1)$$

Thus, the summation of currents at a junction of an electric circuit is analogous to the summation of forces in a
mechanical circuit or to the summation of fluid flow rates
in a fluid circuit. The fact that the algebraic sum of
each of these variables at a junction is zero, is simply
an application of the basic ideas of continuity and equilibrium. Since the through variable in mechanical systems
is force Kirchhoff's current law can be recognized as being
analogous to D'Alembert's principle.

The other circuit law, Kirchhoff's voltage law (KVL),
refers to conditions around a closed path in the circuit
and is:

$$\Sigma(\text{Across Variables}) = 0 \qquad (4.2)$$

Thus, the summation of voltage drops in an electrical
circuit is analogous to the summation of pressure drops
in a fluid circuit or the summation of velocity differences
in mechanical circuits. The summation of these potentials
is sometimes referred to as the compatibility relation.

Although both Kirchhoff's laws are usually stated
(Equations (4.1) and (4.2)) as the algebraic sum being
zero it is often simpler to equate one of the variables
to the sum of the others. For example, the flow into a
junction may be set equal to the flow out of the junction.

When the basic circuit laws are used in this way it is usually possible to set up the equilibrium equations in a form which is suitable for direct solution. The generalization of this process for complex circuits is considered in Chapter VII. However, for simple series, parallel and series-parallel circuits we may apply the basic circuit laws without formal procedures.

Example 4.5

Figure 4.9 shows a series RL electrical circuit with a voltage source, e_S.
   (a) Find the differential equation for the current in terms of the components and voltage source.
   (b) Find a differential equation to represent the voltage across the inductor.
   (c) Find an analogous mechanical circuit. Express the force and velocity in differential equation form.

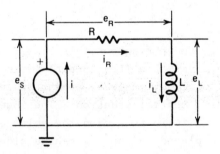

Figure 4.9 Electrical circuit (example 4.5)

   (a) When a current equation is desired always begin with the voltage relation. Thus from KVL around the closed circuit path:

$$- e_s + e_R + e_L = 0$$

The next step is to apply the component relations:

$$- e_s + Ri_R + L \frac{di_L}{dt} = 0$$

In a series circuit, there is only one current. This can be confirmed by applying KCL at any junction. For example, at the junction between the resistor and inductor the KCL yields:

$$- i_R + i_L = 0 \quad \text{or} \quad i_R = i_L$$

where we have used the convention that flows entering a junction are considered negative and flows leaving a junction are considered positive. This is an arbitrary selection. The reverse convention would result in the same relations. In a similar way we could find that $i = i_R$. Since all the currents are the same in the series circuit the differential equation above may be expressed in the form:

$$\frac{L}{R} \frac{di}{dt} + i = \frac{1}{R} e_s$$

This format is used throughout this text.

The dependent variable (the output) is in this case the current, i. It will always appear on the left-hand side of the differential equation. The independent variable (the input) is in this case the voltage source and will always be placed on the right-hand side of the equation.

(b) To determine the voltage, e_L, begin with the current relation at the junction between resistance and inductance, $i_R = i_L$. Then employ the component relations so that:

$$\frac{e_R}{R} = \frac{1}{L} \int e_L \, dt$$

Now use the voltage relation, $e_R + e_L = e_s$ to eliminate e_R. The result is:

$$\frac{e_s - e_L}{R} = \frac{1}{L} \int e_L \, dt$$

Now differentiate and rearrange the equation into the standard format:

$$\frac{L}{R} \frac{de_L}{dt} + e_L = \frac{L}{R} \frac{de_s}{dt}$$

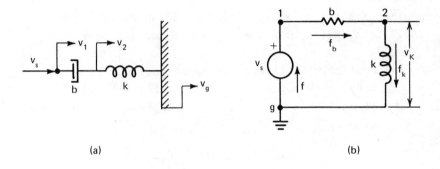

(a)	(b)

Figure 4.10 Analogous mechanical circuit (example 4.5)

(c) Figure 4.10a shows the mechanical arrangement that has the same analogous circuit. It consists of a damper and spring in tandem. A velocity source, v_s, is applied to the free end of the damper. The analogous circuit diagram is shown in Figure 4.10b. Thus:

$$f_b - f_K = 0 \quad \text{(KCLA)}$$

$$-v_1 + (v_1 - v_2) + v_2 = 0$$

$$-v_s + v_b + v_k = 0 \quad \text{(KVLA)}$$

where the designations KCLA and KVLA refer to the analogs
of the basic circuit laws. The substitution of the elemental equations yields:

$$\frac{b}{k}\frac{df}{dt} + f = b\,v_s$$

and

$$\frac{b}{k}\frac{dv_K}{dt} + v_k = \frac{b}{k}\frac{dv_s}{dt}$$

Example 4.6

The parallel RC electrical circuit shown in Figure
4.11 is supplied with energy from a current source, i_s.
Find:

(a) A differential equation representing the voltage
   across the capacitor in terms of R, C, and i_s.

(b) An analogous fluid circuit. Express the pressure
   as a differential equation.

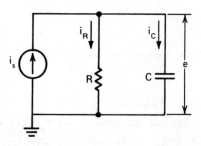

Figure 4.11 Electrical circuit (example 4.6)

(a) Since we are interested in a differential equation
for voltage we begin with the basic current law

$$i_c + i_R = i_s \qquad \text{(KCL)}$$

Substituting the elemental equations gives:

$$C\frac{de}{dt} + \frac{1}{R}\,e = i_s$$

or in standard form:

$$RC \frac{de}{dt} + e = R \ i_s$$

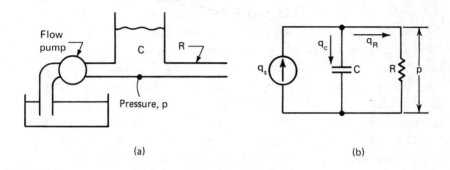

(a)   (b)

Figure 4.12 Analogous fluid circuit (example 4.6)

(b) Figure 4.12a shows the fluid component arrangement that produces the same analogous circuit. This circuit is illustrated in Figure 4.12b. The pressure, p, is the one that exists at the bottom of the tank. The line from the tank is modelled in this case, by a pure resistance element. The procedure for determining the pressure, p, is exactly the same as the procedure for finding, e, in the electrical circuit. Thus, we may write at once that:

$$q_c + q_R = q_s \qquad \text{(KCLA)}$$

and with the use of the component equations

$$RC \frac{dp}{dt} + p = R \ q_s$$

4.3 PROPERTIES OF STORAGE ELEMENTS

Any physical system with energy storage elements requires a differential equation to describe its equilibrium. To predict the response of the system we must find a solution of the differential equation. As shown in Chapters V and VI, this necessitates a foreknowledge of the initial con-

ditions. These initial conditions are not arbitrary mathematical constraints but are rather conditions which result
from the interaction between each energy storage element
and the rest of the circuit at the time of any sudden change
in input. Many possible input functions were considered in
Chapter II. However, several of these were derived from
the basic step function. Accordingly, we now proceed to
investigate the properties of the storage elements when
they are subjected to a sudden change. The properties to
be considered now are the initial response $y(0^+)$ (that is
the response immediately after the change has been applied)
and the final response, $y(\infty)$ (after all transient conditions have disappeared). The complete response, $y(t)$ includes the transient period. This will be determined in
Chapters V and VI.

Initial Conditions

T-Type Elements

 The elemental relationships for T-type storage elements
are given by Equations (3.9), (3.21) and (3.40) for the
electric, fluid, and mechanical systems respectively.
There is no corresponding element in thermal systems.
These elemental equations give the relation between the
across variable (AV) and the through variable (TV) as:

$$AV(t) = L \frac{d}{dt} [TV(t)] \qquad (4.3)$$

The initial response of these elements is more easily
explained, however, by considering the inverse relation:

$$TV(t) = \frac{1}{L} \int_{-\infty}^{t} AV(t')dt' \qquad (4.4)$$

Equation (4.4) is valid, of course, for all values of t.
However, the initial response depends only on the value of
the integral over the infinitesimally small interval from

just before the driving function is applied (t = 0⁻) to
just after it is applied (t = 0⁺). As a result the change
in the value of the through variable at t = 0 is given by

$$\Delta TV(0) = \frac{1}{L} \int_{0^-}^{0^+} AV(t)\,dt \qquad (4.5)$$

If the conditions in the system or circuit are such that
the across variable is finite, the value of this integral
must always be zero. Therefore, the basic property of the
T-type element may be expressed as:

$$TV(0^+) = TV(0^-) \qquad (4.6)$$

If there is no energy stored in the element just before
t = 0 (that is TV(0⁻) = 0) the element provides the constraints that the through variable remains zero as the
driving function is being applied. Thus, when there is
no energy initially stored in a T-type element we may
write:

$$TV(0^+) = 0 \qquad (4.7)$$

In the case of an inductor this means that the inductor
has the effect of keeping the current zero over the interval
from t = 0⁻ to t = 0⁺. As a result the inductor may be
considered to be acting at t = 0 as an open circuit. The
process of finding the initial response of any circuit in
which an inductor (without initial energy storage) is present can therefore be simplified by redrawing the circuit
diagram with the inductor replaced by an open circuit.

The analogous property of inertance in a fluid circuit
constrains the flow rate to remain unchanged in the interval
from t = 0⁻ to t = 0⁺. Thus inertance can also be replaced
at t = 0 by an open circuit. Similarly, the force or torque
in translational or rotational springs cannot change suddenly at t = 0 and they, too, behave initially as open circuits.

Example 4.7

A series RL circuit (Figure 4.13a) is subjected to a step input $e_S = 10u_S(t)$, volts. Find $i(0^+)$, $e(0^+)$, $\frac{di}{dt}(0^+)$.

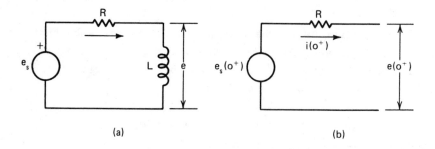

(a)   (b)

Figure 4.13 Initial conditions (example 4.7)

Initially the circuit behaves as if the inductor is an open circuit. This condition is shown by the circuit in Figure 4.13b. We may refer to this circuit as the initial condition circuit.

The current through the initial condition circuit is obviously zero. Thus, $i(0^+) = 0$. The voltage $e(0^+) = e_S(0^+) = 10$ V. To find:

$$\frac{di}{dt}(0^+)$$

we use the elemental inductor equation, $e = L\, di/dt$. The elemental equations hold for all times. We may adapt the equation for $t = 0^+$ as:

$$e(0^+) = L\frac{di}{dt}(0^+)$$

from which

$$\frac{di}{dt}(0^+) = \frac{e(0^+)}{L} = \frac{10}{L}$$

A-Type Elements

The elemental relationships for A-type storage elements
are given in Equations (3.13), (3.25), (3.50) and (3.58)
for electrical, fluid, mechanical, and thermal systems
respectively. All these relationships have the form:

$$AV(t) = \frac{1}{C} \int_{-\infty}^{t} TV(t')dt' \qquad (4.8)$$

The initial response of the A-type elements depends only
on the value of this integral over the infinitesimally
small interval from $t = 0^-$ to $t = 0^+$. Thus, the change
in the value of the across variable at $t = 0$ is given by:

$$\Delta AV(0) = \frac{1}{C} \int_{0^-}^{0^+} TV(t)dt \qquad (4.9)$$

If the conditions in the system or circuit are such
that the through variable is finite the value of this
integral must always be zero. The basic initial property
of the A-type element may therefore be expressed as:

$$AV(0^+) = AV(0^-) \qquad (4.10)$$

When there is no initial energy storage in the element
($AV(0^-) = 0$), the across variable remains zero as the
driving function is being applied. Thus, we may write:

$$AV(0^+) = 0 \qquad (4.11)$$

For the case of an electrical capacitor (without initial
charge) the voltage drop across it remains equal to zero
in the interval from $t = 0^-$ to $t = 0^+$. Therefore, the
electrical capacitor may be considered to be acting at
$t = 0^+$ as a short circuit. To find the initial response
of a circuit with an electrical capacitor can be simplified
by redrawing the circuit diagram with the capacitor replaced
by a short circuit.

The analogous property of a fluid capacitor keeps the
pressure constant in the interval from t = 0⁻ to t = 0⁺.
This is represented on the analogous circuit diagram by
replacing the fluid capacitor with a short circuit at
t = 0⁺. Similarly, the mass and moment of inertia in
translational and rotational mechanical systems do not
permit any sudden change in velocity or angular velocity.
In addition thermal mass inhibits an instantaneous change
in temperature. All of these components behave as short
circuits at t = 0⁺.

Example 4.8

A mass is connected to a damper as shown in Figure
4.14a. The mass is initially at rest $(v(0^-) = 0)$. A
step input of force, $f_s = 5u_s(t)$, N is applied to the
mass. Find $v(0^+)$ and $dv/dt(0^+)$.

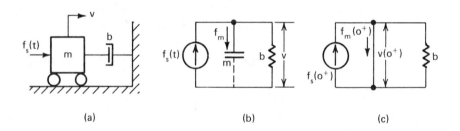

(a)   (b)   (c)

Figure 4.14 Initial conditions (example 4.8)

The analogous circuit for the mechanical arrangement
is shown in Figure 4.14b. The circuit has an across
type storage device and a dissipative element in parallel.
The input is represented by an ideal force source. We
may convert the analogous circuit to an initial condition
circuit by shorting out the A-type storage device. Figure
4.14c shows the resulting initial condition circuit for
this case. To determine the required values $v(0^+)$ and
$dv/dt(0^+)$, we analyze the initial condition circuit.

Evidently, $v(0^+) = 0$, since the analogous circuit behaves as if the mass (capacitor) is a short circuit to ground. The relationship

$$f_m = m \frac{dv}{dt}$$

is the equation for the mass element and holds at every instant of time. For the specific case of $t = 0^+$ we find that:

$$\frac{dv}{dt}(0^+) = \frac{f_m(0^+)}{m}$$

Thus, the problem of determining $dv/dt(0^+)$ is reduced to the simpler problem of finding $f_m(0^+)$. Initial condition circuit (Figure 4.14c) indicates that the force taken initially by the mass ($f_m(0^+)$) is equal to the entire force supplied at $t = 0$ that is $f_m(0^+) = f_s(0^+) = 5$. As a result:

$$\frac{dv}{dt}(0^+) = \frac{5}{m}$$

Remember that although the mass is behaving as a short-circuit, the mass element is still located between the velocity terminal and ground. Thus, all the force from the ideal source "flows" through the mass at $t = 0^+$.

Final Conditions With Step Inputs

We have observed that the initial values of the response to a sudden change such as a step input can be obtained directly from a circuit representation that is valid only at $t = 0^+$. In a similar way when the input is a step we may obtain the final values from a circuit equivalent that is valid only at $t = \infty$. As we shall see in Chapters V and VI, the final values provide the particular solutions for the circuit differential equations when a step input is applied.

One of the main characteristics of any linear system is that the particular solution is the part of the complete solution that is due to the forcing or driving function (i.e., the input). The particular solution must have the same shape as the forcing function. The only differences between forced response (output) and driving function (input) shape are a scale factor and depending on the driving function, a timing factor. For a step input the forced response is a constant after the step is applied. Therefore, there is no timing factor. On the other hand, the forced response to a sine wave input has both a scale factor (magnitude) and a timing factor (phase difference).

The scale factor for a step input is obtained by noting that the shape of the input is not changing after $t = 0^+$. The forced response of a linear system is, therefore, also characterized by an unchanging value.

One way to derive the final values of circuit variables is to determine a final condition circuit. As in the initial condition circuit this requires the replacement of the storage elements with equivalent open and short circuits.

T-Type Elements

The final state of a T-type element under the conditions imposed by step excitation is more easily derived by returning to the general elemental relationship given in Equation (4.3). In this case, we note that the final condition of a step is characterized by zero rate of change. For a linear system, this zero rate of change must also be a characteristic of the final state of all the elements within the system. Thus Equation (4.3) may be rewritten at $t = \infty$ as:

$$AV(\infty) = 0 \qquad\qquad (4.12)$$

In the case of an inductor this means that the final value of the voltage across it is zero, and it may therefore be represented by a short-circuit. This, of course, applies

only to the ideal inductor which it should be recalled,
has no resistance. Similarly, the final pressure difference associated with inertance is zero, and the final
difference in velocity between the ends of a spring is
zero. The process of finding the final state of a system
or circuit in which T-type elements are present is therefore quite simple if the circuit diagram is redrawn with
T-type elements replaced by short-circuits.

A-Type Elements

The final state of an A-type element when it is part of
a system subjected to a step in excitation is determined
by considering the inverse form of Equation (4.8):

$$TV(t) = C \frac{d}{dt} [AV(t)] \qquad (4.13)$$

Again, we note that with a step function the final rate
of change is zero, and therefore the final value of the
through variable associated with an A-type element is:

$$TV(\infty) = 0 \qquad (4.14)$$

In the case of an electrical capacitor this means that
there is no flow of current through it when the voltage
across it is constant. This situation is simply represented
on a circuit diagram by redrawing it with the capacitor
replaced by an open circuit. Similarly, the final flow rate
into a fluid capacitor is zero, the final force or torque
acting on a mass or inertia is zero, and the final flow of
heat into thermal capacitance is zero. For the final state
with step excitation all of these situations are easily
considered by redrawing the analogous circuit diagram with
the A-type elements replaced by open circuits.

Example 4.9

Figure 4.15a shows a mechanical model that consists of
two masses, m_1 and m_2, two dampers, b_1 and b_2, and one
spring, k. A step function

$$f_s(t) = 2u_s(t), \text{ N}$$

is applied to mass, m_1. The velocity of mass, m_1 is v_1 and the velocity of mass, m_2, is v_2. Find

(a) $v_1(0^+)$, $v_2(0^+)$, $\dfrac{dv_1}{dt}(0^+)$, $\dfrac{dv_2}{dt}(0^+)$

(b) $v_1(\infty)$, $v_2(\infty)$

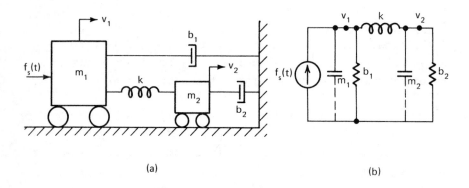

(a)

(b)

Figure 4.15   Mechanical circuit (example 4.9)

Figure 4.15b shows the analogous circuit representation of the mechanical model. There are two terminals v_1, and v_2, plus a ground. Notice that the mass components are always between a terminal and ground. Once the terminals have been numbered and the masses have been positioned it is a simple matter to connect the other components. Thus, the spring connects the two mass terminals together and the dampers go from a mass to ground. To determine the initial and final conditions sought, it is convenient to draw initial and final condition circuits. These circuits contain only dissipative elements and sources.

(a) Figure 4.16a shows the initial condition circuit. This circuit is obtained by replacing the A-type storage devices m_1 and m_2 by short-circuits and replacing the T-type storage device, k, by an open circuit. The dissi-

pative elements, b_1 and b_2 remain in place. Now we analyze the initial condition circuit to determine the
initial values of the variables. Since the masses are
represented by short-circuits to ground $v_1(0^+)$ and $v_2(0^+)$
have the same velocity as ground. Thus $v_1(0^+) = v_2(0^+)$
$= 0$. To determine the initial values of the derivatives
we must find a component in the original circuit (Figure
4.15b) that has an elemental equation containing the
derivative. Thus from m_1 we know that:

$$f_{m_1} = m_1 \frac{dv_1}{dt}$$

and from m_2 that:

$$f_{m_2} = m_2 \frac{dv_2}{dt}$$

These equations are valid for all times. Thus, for the
special case of interest here, we may write that:

$$\frac{dv_1}{dt}(0^+) = \frac{f_{m_1}(0^+)}{m_1} = \frac{2}{m_1}$$

where

$$f_{m_1}(0^+) = f_s(0^+) = 2$$

is obtained by noting the initial condition circuit
behavior. In addition

$$\frac{dv_2}{dt}(0^+) = \frac{f_{m_2}(0^+)}{m_2} = 0$$

since there is no source in the loop of the initial
condition circuit that encompasses m_2.

(b) The final conditions can be obtained by drawing a
circuit equivalent to $t = \infty$ as shown in Figure 4.16b.
As in the case of the initial condition circuit the
elements in the final condition circuit are all dissi-

pative and sources. However, in this circuit the masses
behave like open circuits and the spring behaves like a
short circuit. It is evident from the resulting final
condition circuit that the two masses must have the same
velocity and that the dampers are effectively in parallel.
Thus $f_S(\infty) = (b_1 + b_2)v(\infty)$ or

$$v_1(\infty) = v_2(\infty) = \frac{2}{b_1 + b_2}$$

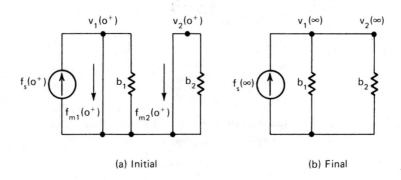

(a) Initial   (b) Final

Figure 4.16 Initial and final condition circuits (example 4.9)

A word of caution is required regarding the use of the
designation "parallel". Its use in this example to describe the constraints on the dampers b_1 and b_2 is
limited to the final state where the difference in
velocity across the spring is zero. At all other times,
the two masses have different velocities. Thus, these
two dampers must not be considered as being in parallel
as far as the general problem of describing the system
behavior is concerned. It must be remembered that before
any two elements in any system may be considered to be
in parallel, the circuit representation must show a
definite constraint of equal across variables.

Example 4.10

The electrical circuit shown in Figure 4.17 has R_1 = 4 ohms, R_2 = 3 ohms, C = 0.2 F and L = 1.0 H. If a current source $i(t) = 2u_s(t)$, A, find:

(a) $e_1(0^+)$, $e_2(0^+)$, $e_3(0^+)$, $\dfrac{de_3}{dt}(0^+)$, $\dfrac{di_2}{dt}(0^+)$

(b) $e_1(\infty)$, $e_2(\infty)$, $e_3(\infty)$, $i_1(\infty)$, $i_2(\infty)$

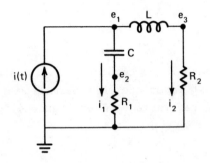

Figure 4.17   Electrical circuit (example 4.10)

(a) To find the initial conditions (t = 0^+) we make an initial condition circuit. This is shown in Figure 4.18a. From the initial condition circuit we observe that $i_1(0^+)$ = 2A and $i_2(0^+)$ = 0. Thus:

$$e_1(0^+) = e_2(0^+) = R_1\ i_1(0^+) = 8\ \text{V}$$
$$e_3(0^+) = R_2\ i_2(0^+) = 0\ \text{V}$$

To evaluate

$$\frac{di_2}{dt}(0^+)$$

we seek a component that has an elemental equation containing this term. Note from Figure 4.17 that:

$$e_1 - e_3 = L \frac{di_2}{dt}$$

This equation holds for all time. For the special case of $t = 0^+$ therefore:

$$\frac{di_2}{dt}(0^+) = \frac{1}{L}[e_1(0^+) - e_3(0^+)] = 8.0 \text{ A/s}$$

There remains now to determine

$$\frac{de_3}{dt}(0^+)$$

In this case there is no component that contains this term. Thus we must formulate this term from another component by differentiation. From Figure 4.17:

$$e_3 = R_2 i_2$$

and by differentiation:

$$\frac{de_3}{dt} = R_2 \frac{di_2}{dt}$$

This relationship also must hold for all times. At $t = 0^+$

$$\frac{de_3}{dt}(0^+) = R_2 \frac{di_2}{dt}(0^+) = 24.0 \text{ V/s}$$

(b) Figure 4.18b shows the final condition circuit representation for this problem. From the final condition circuit we see that $i_1(\infty) = 0$ and $i_2(\infty) = 2.0$ A. Thus:

$$e_1(\infty) = e_3(\infty) = R_2 i_2(\infty) = 6.0 \text{ V}$$
$$e_2(\infty) = R_1 i_1(\infty) = 0 \text{ V}$$

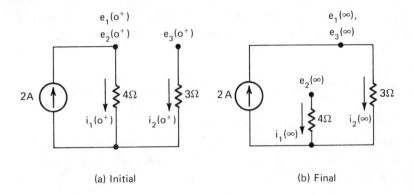

(a) Initial   (b) Final

Figure 4.18 Initial and final condition circuits (example 4.10)

It must be emphasized that the foregoing method for determining the final conditions is applicable only when the input is a step function.

4.4 THE MATHEMATICAL APPROACH TO INITIAL CONDITIONS

There is a mathematical method to determine the initial conditions in a circuit when the input function is any arbitrary time function. We will discuss this method here only briefly. However, this should be sufficient to permit us to check the initial condition results obtained from the initial condition circuit.

To use the mathematical approach we must first find the differential equation for the variable whose initial value we wish to know as dependent variable. Suppose we designate the variable as $y(t)$ and the governing equation as:

$$\frac{d^2y}{dt^2} + \frac{dy}{dt} + y = a_1 g(t) + a_2 \frac{dg}{dt}(t) \qquad (4.15)$$

where $g(t)$ is the input function and a_1 and a_2 are constants. We want to find $y(0^+)$ and $dy/dt(0^+)$. The procedure is to

integrate Equation (4.15) until the left-hand side has only
one non-integral term. After one integration, Equation
(4.15) becomes:

$$\frac{dy}{dt} + y + \int_{-\infty}^{t} y(t')dt' = a_1 \int_{-\infty}^{t} g(t')dt' + a_2\, g(t) \quad (4.16)$$

where t' is a variable of integration.

There are two non-integral terms on the left-hand side
of Equation (4.16). Thus, we must perform another integration to reach:

$$y(t) + \int_{-\infty}^{t} y(t')dt' + \int_{-\infty}^{t}\int_{-\infty}^{t'} y(t'')dt''dt'$$

$$= a_1 \int_{-\infty}^{t}\int_{-\infty}^{t'} g(t'')dt''dt' + a_2 \int_{-\infty}^{t} g(t')dt' \quad (4.17)$$

where t" is also a variable of integration. Now, remember
that Equation (4.17) is valid for all times. Thus we may
evaluate Equation (4.17) at t = 0$^+$ directly if the second
term on the right-hand side is not an impulsive function
(i.e., the impulse, doublet, etc.). With this condition
the evaluation of Equation (4.17) at t = 0$^+$ yields:

$$y(0^+) = a_2 \int_{-\infty}^{0^+} g(t')dt' \quad (4.18)$$

where the other integral terms in Equation (4.17) are all
identically equal to zero. This is a consequence of integrating non-impulsive functions between the limits -∞ to
0$^+$. The initial condition of the first derivative can
now be obtained by evaluation of Equation (4.16) at t = 0$^+$
and using Equation (4.18). The result is:

$$\frac{dy}{dt}(0^+) = (a_1 - a_2) \int_{-\infty}^{0^+} g(t')dt' + a_2\, g(0^+) \quad (4.19)$$

Let us try out this procedure on a few simple circuits.

Example 4.11

Figure 4.19a shows the schematic drawing of a pneumatic circuit in which a pump supplies a tank through a length of line. When the pump is suddenly turned on it develops the pressure function $p(t) = u_s(t)$. If the line can be modelled by a pure resistance, find the flow into the tank immediately after the pump is turned on.

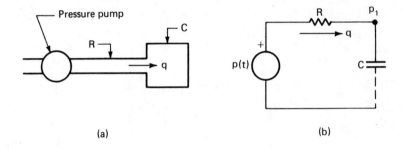

(a)   (b)

Figure 4.19   Fluid circuit (example 4.11)

The first step of the mathematical approach is to determine the differential equation that describes the flow, $q(t)$. To do this, consider the analogous circuit diagram shown in Figure 4.19b. By KVLA we know that:

$$[p(t) - p_1] + [p_1] = p(t)$$

and by substituting the circuit component equations we obtain:

$$Rq + \frac{1}{C} \int q \, dt = p(t)$$

and by differentiation

$$RC \frac{dq(t)}{dt} + q(t) = C \frac{dp(t)}{dt} = Cu_i(t)$$

Now according to the mathematical procedure we must use repeated integration until there is only one non-integral term on the left-hand side.   In this case after one integration

$$RC\ q(t) + \int_{-\infty}^{t} q(t')dt' = Cu_s(t)$$

and we have already reached the condition of only one non-integral term on the left-hand side.   We may therefore evaluate the equation at $t = 0^+$ with the result that:

$$RC\ q(0^+) = C \quad \text{and} \quad q(0^+) = \frac{1}{R}$$

since

$$\int_{-\infty}^{t} q(t')dt' = 0 \quad \text{when } t = 0^+$$

Of course, we would have reached this same result by simply replacing the capacitor in Figure 4.19b by a short-circuit.   Then the flow through the resulting initial condition circuit would have been $q(0^+) = 1/R$ directly.

Example 4.12

The mechanical mass-spring-damper model shown in Figure 4.20a is subjected to a velocity input

$$v_s(t) = u_s(t)$$

Find the initial value of the force taken by the spring $(f_k(0^+))$ and the initial value of the time derivative of this force $(df_k/dt(0^+))$.

Figure 4.20b shows the analogous circuit representation of the model.   Since we are interested in the spring force we must derive the differential equation for this vari-

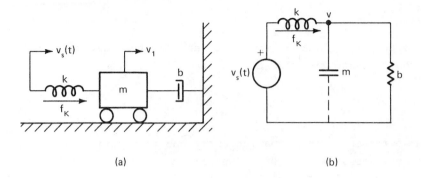

Figure 4.20 Mechanical circuit (example 4.12)

able. By use of the circuit laws and some algebra we may express the spring force as:

$$\frac{d^2 f_k}{dt^2} + \frac{b}{m} \frac{df_k}{dt} + \frac{k}{m} f_k = k \frac{dv_s}{dt} + \frac{bk}{m} v_s$$

If we substitute $v_s = u_s(t)$ into the above equation, the result is:

$$\frac{d^2 f_k}{dt^2} + \frac{b}{m} \frac{df_k}{dt} + \frac{k}{m} f_k = k u_i(t) + \frac{bk}{m} u_s(t)$$

The mathematical approach requires integration until only one non-integral term remains on the left. One integration produces:

$$\frac{df_k}{dt} + \frac{b}{m} f_k + \frac{k}{m} \int_{-\infty}^{t} f_k \, dt' = k u_s(t) + \frac{bk}{m} u_r(t)$$

A second integration leads to:

$$f_k + \frac{b}{m} \int_{-\infty}^{t} f_k \, dt' + \frac{k}{m} \int_{-\infty}^{t} \int_{-\infty}^{t'} f_k \, dt''dt'$$

$$= k u_r(t) + \frac{bk}{m} u_p(t)$$

Now evaluation of the above result at $t = 0^+$ yields:

$$f_k(0^+) = 0$$

since the integral terms are zero and the ramp and parabolic terms are also zero at $t = 0^+$. To find

$$\frac{df_k}{dt}(0^+)$$

we return to the equation after one integration and
evaluate it at $t = 0^+$. The result is:

$$\frac{df_k}{dt}(0^+) = k$$

since we know that

$$f_k(0^+) = 0$$

and the ramp function is zero at $t = 0^+$.

In the initial condition circuit method the spring
behaves like an open circuit. Thus $f_k(0^+)$ is immediately known as zero. The equation of the spring is:

$$\frac{df_k}{dt} = k(v_s - v_1)$$

at $t = 0^+$, $v_1(0^+) = 0$ because of the shorting of the
mass. Thus:

$$\frac{df_k}{dt}(0^+) = k$$

by this method also.

If impulsive functions remain on the right-hand side
after the appropriate number of integrations the evaluation
at $t = 0^+$ cannot be performed since the left-side integrals
are not zero. In this circumstance, additional integrations
are required to remove the impulsive functions. The sub-

sequent evaluation at $t = 0^+$ then defines an integral of the variable, i.e.

$$\int_{-\infty}^{0^+} y(t')dt', \quad \int_{-\infty}^{0^+} \int_{-\infty}^{0^+} y(t'')dt''dt'$$

The initial conditions are then determined from this value by working back through successive integrations.

4.5 CIRCUIT MODELS FOR INITIAL ENERGY STORAGE

In Section 4.3, the energy storage elements had no stored energy prior to the application of the step function. With this restriction the A-type storage element behaves initially as a short circuit and the T-type storage element behaves initially as an open circuit.

The simplest way to treat storage elements with initial energy storage is to introduce a special circuit representation for such cases. Since these circuits must be based on physical equations, let us begin with the mathematical formulation of energy storage.

A-Type Storage

Consider the elemental equation of the A-type storage device in similar form to Equation (4.8)

$$AV(t) = \frac{1}{C} \int_{-\infty}^{t} (TV)dt' \qquad (4.20)$$

The use of the lower limit of integration as $t = -\infty$ includes the entire previous history of the storage device. We can separate the integral in Equation (4.20) into two integrals. Thus:

$$AV(t) = \frac{1}{C} \int_{-\infty}^{0} (TV)dt' + \frac{1}{C} \int_{0}^{t} (TV)dt' \qquad (4.21)$$

We can interpret the first integral as the value of the across variable at $t = 0$ (i.e. $AV(0)$). Hence, for all

values of time greater than zero, the elemental equation
is:

$$AV(t) = AV(0) \, u_s(t) + \frac{1}{C} \int_0^t (TV) \, dt' \qquad (4.22)$$

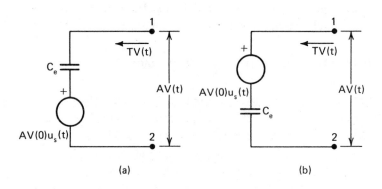

(a)   (b)

Figure 4.21 Circuit models for A-Type storage

The first term on the right-hand side represents the
value of the across variable due to initial energy storage.
The second term is recognized as the standard expression
for an A-type storage device. Now a circuit model or
equivalent circuit must yield the same relation between
$AV(t)$ and $TV(t)$ as indicated in Equation (4.22). This
relation suggests a series connection of a source (to
represent $AV(0) \, u_s(t)$) and an A-type storage element.
These components may be placed in either order. The
corresponding circuit models are shown in Figure 4.21.
Each model has identically the same equilibrium equation
as Equation (4.22) provided that $C = C_e$. We, therefore,
conclude that they are equivalent circuits that may be
used to replace A-type elements with initial storage.
The reason for designating the component in Figure 4.21
as C_e is to emphasize that this component alone does not
correspond to the original component. Since this is a
true equivalent circuit the terminals representing the
actual component are marked as 1 and 2 on Figure 4.21.

Example 4.13

A tank filled with water is emptied through a long pipe as indicated in Figure 4.22a. If the initial height of the water is h_o, draw an equivalent circuit after the valve is opened. Locate the terminal on the circuit drawing that represents the pressure at the bottom of the tank.

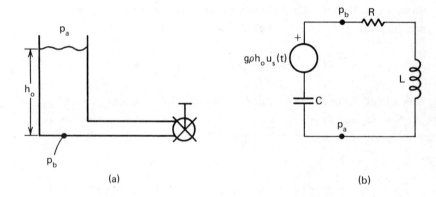

Figure 4.22 Fluid circuit (example 4.13)

The initially filled tank is a fluid capacitor with initial storage. We may choose either circuit model shown in Figure 4.21 to represent this component. However, do not forget that one terminal of the equivalent circuit for a tank capacitor must always go to ground. The initial storage pressure is ρgh_o (where ρ is the water density and g is the gravitational constant). Thus, the tank model has a source, $\rho gh_o\ u_s(t)$, in series with a capacitance. The general line model is an R and L in series. Figure 4.22b shows the complete analogous circuit for the fluid arrangement in Figure 4.22a. The pressure at the bottom of the tank, p_b, is shown on the analogous circuit. Note that this terminal does not come between the source and capacitor. There is no available terminal between these components.

T-Type Storage

The corresponding model for the T-type storage element is obtained by performing the dual analysis. We begin with the basic elemental equation (Equation (4.3)) in integral form. Thus:

$$TV(t) = \frac{1}{L} \int_{-\infty}^{t} (AV)dt' \qquad (4.23)$$

When expanded in the same manner as for the A-type element Equation (4.23) becomes:

$$TV(t) = \frac{1}{L} \int_{-\infty}^{o} (AV)dt' + \frac{1}{L} \int_{o}^{t} (AV)dt' \qquad (4.24)$$

The first integral is the value of the through variable at $t = 0$ (i.e. $TV(0)$). For times greater than zero we may express the elemental equation given in Equation (4.24) as:

$$TV(t) = TV(0) \ u_s(t) + \frac{1}{L} \int_{o}^{t} (AV)dt' \qquad (4.25)$$

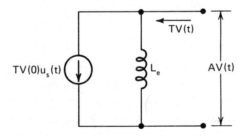

Figure 4.23 Circuit model for T-Type storage

Now the first term on the right-hand side of Equation (4.25) is the value of the through variable due to initial storage. The integral term is the standard relation for a T-Type storage element. To form a circuit model from Equation (4.25) note that each term represents a through

variable. The sum of through variable terms must result
in a parallel connection according to our basic circuit
laws. Figure 4.23 shows the circuit that produces the
relation between TV(t) and AV(t) given in Equation (4.25)
provided that $L_e = L$. Thus, a T-type storage element is
modelled by an equivalent element (having the same numerical
parameter value) in parallel with a step flow source.

Example 4.14

The spring-damper system shown in Figure 4.24a is given
an initial displacement, X_0, and then released. Obtain
an equivalent circuit for the system and indicate the
branch that carries the spring force, f_k.

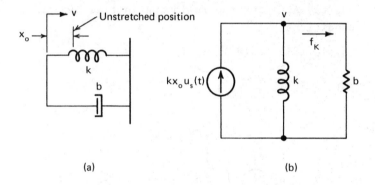

Figure 4.24 Mechanical circuit (example 4.14)

The spring in the mechanical arrangement is modelled
by the parallel combination of a force source, $k\, X_0$
$u_s(t)$, and a spring component. The damper is also connected between the velocity terminal, v, and ground.
Figure 4.24b shows the analogous circuit diagram. Note
that the initial displacement produces a reaction force
that is in the direction of the assumed velocity. This
force is modelled, therefore, as directed toward the
node v. The spring force, f_k, is located also on Figure

4.24b. Note that this is not the through variable
passing through the k element in the circuit. The spring
force is the sum of two force terms (the source contribution and the element contribution). This is a consequence of the circuit model for T-type storage.

The complete mathematical solutions for physical circuits
with or without initial storage are discussed in Chapters
V and VI.

Problems

4.1 For the electric circuit shown in Figure P4.1,
R_1 = 2 Ω, R_2 = 5 Ω, C = 5 μF and L = 0.1 H.   The
current generator produces the step input i(t) =
$5u_s(t)$, A.   Find the following initial and final
conditions:

(a) $i_{R_1}(0^+)$, $e_2(0^+)$, $i_{R_1}(\infty)$, $e_2(\infty)$

(b) $\dfrac{de_2}{dt}(0^+)$, $\dfrac{di_L}{dt}(0^+)$

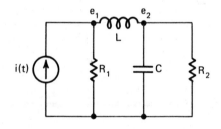

Figure P 4.1

4.2 Some initial and final conditions are desired for
the electric circuit shown in Figure P4.2.   Find:

(a) $e_1(0^+)$, $e_2(\infty)$, $i_{R_1}(0^+)$, $i_{R_1}(\infty)$, $i_{R_3}(\infty)$

(b) $\dfrac{de_2}{dt}(0^+)$, $\dfrac{de_1}{dt}(0^+)$

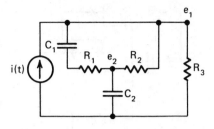

Figure P 4.2

4.3 The rotational mechanical circuit shown in Figure
P4.3 has $J = 100$ kg·m^2, $K = 20$ N·m, and $B = 5$ N·m·s.
If the input torque $T(t) = 500u_s(t)$, N·m, find:

(a) $\omega_J(0^+)$, $\omega_J(\infty)$, $T_B(\infty)$

(b) $\dfrac{d\omega_J}{dt}(0^+)$

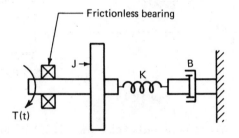

Figure P 4.3

4.4 In the mechanical circuit shown in Figure P4.4,
$b_1 = 1$ N·s/m, $b_2 = 3$ N·s/m, $k = 5$ N/m and $m = 20$ kg.
The input velocity, $v(t) = 20u_s(t)$, m/s. Find:

(a) $f_m(0^+)$, $f_k(\infty)$, $v_m(\infty)$

(b) $\dfrac{dv_m}{dt}(0^+)$, $\dfrac{df_k}{dt}(0^+)$

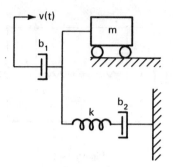

Figure P 4.4

4.5 Draw the electric circuit analog for the mechanical circuit shown in Figure P4.5.

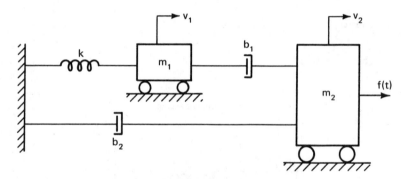

Figure P 4.5

4.6 Find an analogous electrical circuit for the fluid circuit shown in Figure P4.6. If the constant flow pump were replaced by a constant pressure pump which element could be removed without changing the circuit performance?

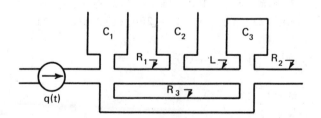

Figure P 4.6

4.7 Draw the electric circuit analog for the rotational
mechanical circuit in Figure P4.7.

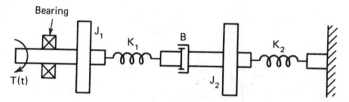

Figure P 4.7

4.8 Find the analogous electrical circuit for the thermal
system given in Figure P4.8. The water is a good
heat conductor and is primarily a storage medium.
The insulators have poor heat storage qualities.
The surface coefficient of heat transfer, h, for
insulation I, is very high.

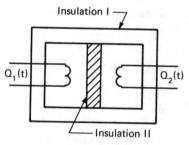

Figure P 4.8

4.9 A fluid system is shown in Figure P4.9. How would
you represent the system by an analogous electrical
circuit?

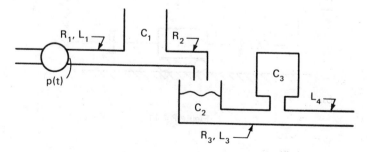

Figure P 4.9

4.10 Find the mechanical arrangement that has the electric
circuit analog shown in Figure P4.10.

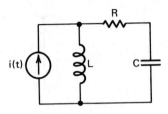

Figure P 4.10

4.11 The mechanical system shown in Figure P4.11 is acted
upon by a force, $f(t) = 10u_s(t)$, N. Initially, the
10 kg mass is moving away from the wall at a velocity
of 1 m/s and initially there is no force in the
spring. The damper $b = 4$ N·s/m and the spring
$k = 1$ N/m. Find:

(a) The initial value of:

$$\frac{df_k}{dt} \quad \text{and} \quad \frac{df_b}{dt}$$

(b) The final value of f_k and v_1.

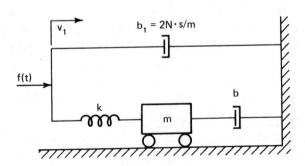

Figure P 4.11

CHAPTER 5

RESPONSE OF FIRST ORDER CIRCUITS

First order circuits normally contain only one storage element and may contain one or more dissipative elements. The storage element may be A-type or T-type. Irrespective of the total number of circuit elements, the circuit with one storage element will always be described by a first order differential equation.

In the first three sections of this Chapter, we consider circuits in which the storage element has no initial energy storage. This condition is sometimes referred to as "initially at rest". In terms of the signal variables initially at rest means that the across variables of A-type storage elements and the through variable of T-type storage elements are equal to zero before the circuit is activated ($t = 0^-$). Circuits with initial storage will be treated in Section 5.4.

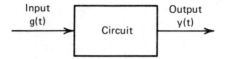

Figure 5.1 Block diagram of circuit

The response of circuits is the time variation that occurs
in a specific signal variable as the result of a time variation in another variable. Figure 5.1 shows this concept in
block diagram form. When the general input g(t) is operated
on by a circuit an output function y(t) is produced. The
output is the response of the circuit to a particular input.
The most general differential equation for first order
circuits is:

$$\tau \frac{dy}{dt} + y = a_1 \, g(t) + a_2 \, \frac{dg(t)}{dt} \qquad (5.1)$$

where τ, a_1, and a_2 are constants that depend on the values
of the circuit elements. The constant, τ, is usually
referred to as the "time constant" of the circuit. The
physical significance of the time constant will become
evident when we obtain solutions for Equation (5.1). For
circuits that have more than one dissipative element a_1 or
a_2 may be zero. However, in the case of a circuit with
precisely one storage element and one dissipative element
either a_1 or a_2 must be zero.

   The following sections of this Chapter will develop the
circuit response, y(t), for a variety of input functions,
g(t).

5.1 STEP RESPONSE

   Let us begin with a two-element circuit - one storage
element and one dissipative element. If the storage element and the output variable associated with it are of the
same type (e.g., a capacitor and the voltage across it or

an inductor and the current through it) then $a_2 = 0$ in Equation (5.1). In addition, the output variable in this type of circuit cannot change suddenly when a step change is made in the input and therefore $y(0^+) = y(0^-) = 0$. We will call this a type 1 circuit. From Equation (5.1) with $a_2 = 0$ the differential equation describing a type 1 circuit is:

$$\tau \frac{dy}{dt} + y = a_1 \ g(t) \qquad\qquad (5.2)$$

$$y(0^+) = 0$$

The input function $g(t) = G \ u_s(t)$ where G is the amplitude of the step change. Thus, Equation (5.2) becomes:

$$\tau \frac{dy}{dt} + y = a_1 \ G \ u_s(t) \qquad\qquad (5.3)$$

We will solve Equation (5.3) for y in the classical mathematical manner as the sum of a homogeneous solution y_H and a particular solution, y_P. In engineering we refer to the homogeneous solution as the natural response and the particular solution as the forced response. Thus:

$$y = \underset{\uparrow\downarrow}{y_H} \qquad + \qquad \underset{\uparrow\downarrow}{y_P} \qquad\qquad (5.4)$$

$$= y_{NATURAL} \quad + \quad y_{FORCED}$$

The homogeneous solution is sometimes also called the unforced solution. The homogeneous differential equation is, therefore, obtained from the complete equation by setting all input functions equal to zero. The homogeneous and particular forms of Equation (5.3) are:

$$\tau \frac{dy_H}{dt} + y_H = 0 \qquad\qquad (5.5a)$$

$$\tau \frac{dy_P}{dt} + y_P = a_1 \ G \ u_s(t) \qquad\qquad (5.5b)$$

Note that the addition of Equations (5.5a) and (5.5b)
yields Equation (5.3). The homogeneous equation (Equation
(5.5a)) has variables which are separable and may be integrated directly to obtain the homogeneous solution. However, we do not use this method of solution here. We
prefer rather a method that is applicable to this and
more complicated circuits. In this method we assume a
trial solution for y_H in the form A e^{rt} where A and r are
constants to be determined. The rationale for this assumption is that the form A e^{rt} can be made to represent all
the expected response functions. Thus, if y_H = A e^{rt} is
substituted into Equation (5.5a) the result is:

$$(\tau r + 1) \text{ A } e^{rt} = 0 \qquad\qquad (5.6)$$

Since the trial solution A $e^{rt} \neq 0$, Equation (5.6) yields
the condition that:

$$\tau r + 1 = 0 \qquad\qquad (5.7)$$

Equation (5.7) is known as the characteristic equation
of the circuit. The equation is satisfied if $r = -1/\tau$.
Thus, the homogeneous solution becomes:

$$y_H = \text{A } e^{-t/\tau} \qquad\qquad (5.8)$$

To obtain the particular solution let us consider Equation
(5.5b). When the input function is a step function the
right-hand side is a constant for t > 0. Thus, the sum
of the left-hand terms must also be a constant. In effect,
we need a function which added to its derivative yields a
constant.

The particular solution that satisfies Equation (5.5b)
is:

$$y_P = a_1 \text{ G } u_s(t) \qquad\qquad (5.9)$$

The correctness of this solution is readily verified by substituting it back into Equation (5.5b). Another way of approaching the particular solution is to note that it is the same as the value of the variable after all the transients have disappeared. Remember that in physical circuits the output is not changing a long time after a step has been applied. This means that $dy/dt = 0$ when t approaches infinity. As a result, the particular solution for a circuit with a step input is equal to the value of the output variable at $t = \infty$. Thus we may also write:

$$y_p = y(\infty) \qquad (5.10)$$

The complete solution of Equation (5.3) is the sum of the homogeneous and particular solutions given in Equations (5.8) and (5.9)

$$y(t) = A\,e^{-t/\tau} + a_1\,G\,u_s(t) \qquad (5.11)$$

Now that the complete solution has been obtained we may evaluate the constant A from the initial condition $y(0^+) = 0$. Thus:

$$y(0) = A + a_1\,G = 0 \qquad (5.12)$$

from which we find that $A = -a_1\,G$. The complete response of this first-order circuit to a step input is therefore:

$$y(t) = a_1\,G\,[1 - e^{-t/\tau}]u_s(t) \qquad (5.13)$$

Figure 5.2 shows the response described by Equation (5.13). The ordinate is the normalized output $y(t)/a_1 G$ and the abscissa is the number of time constants. At $t = \tau$ (one time constant) the normalized output is 0.632 of its final value. After two-time constants ($t = 2\tau$) the output is 0.865 and after three-time constants ($t = 3\tau$) it is 0.950.

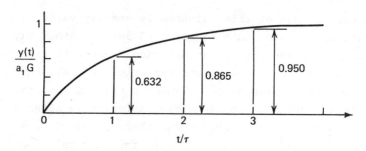

Figure 5.2 Normalized step response of type 1 circuit

We have presented a mathematical treatment of a type 1
circuit. Now let us reinforce some of these ideas with
an example of a particular physical circuit.

Example 5.1

A velocity step input is applied to the mechanical
damper and mass arrangement shown in Figure 5.3a. If
$v(t) = 2u_s(t)$, m/s, $b = 5.0$ N·s/m and $m = 50$ kg, find
$v_m(t)$.

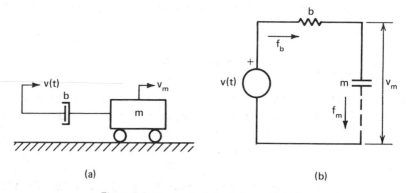

(a)   (b)

Figure 5.3 Mechanical circuit (example 5.1)

The first step is to place the physical arrangement
into a circuit form. This is shown in Figure 5.3b. We
may recognize from both the mechanical arrangement and

the circuit that the force passing through the damper is
applied to the mass. This is a series circuit so that:

$$f_m = f_b$$

Now from the elemental equations we know that
$f_m = m dv_m/dt$ and $f_b = b(v - v_m)$. If these equations
are substituted above we obtain:

$$m \frac{dv_m}{dt} = b(v - v_m)$$

which may be arranged in the form given in Equation (5.2)
for a type 1 circuit

$$(\frac{m}{b}) \frac{dv_m}{dt} + v_m = v(t)$$

where m/b is the circuit time constant and $a_1 = 1$. We
might have expected that this circuit would be of type
1 since the output variable v_m is an across variable and
the mass is an across storage element. For the particular
component values given $\tau = m/b = 50/5 = 10$ seconds,
$G = 2.0$ m/s, the desired solution follows directly from
Equation (5.13) as:

$$v_m(t) = 2(1 - e^{-t/10})u_s(t), \; m/s$$

The mass starts from rest and ultimately reaches the
velocity of the input. After 10 seconds (1 time constant)
the mass is moving at 0.632 of the final velocity or
1.264 m/s.

The solution presented above is based on the mathematical development. Let us try the more physical approach
of adding the natural and forced responses to obtain the
same solution. The physical approach may be summarized
in four steps:

(1) Find the homogeneous solution. For the first order circuit this solution is always

$$Ae^{-t/\tau}$$

For the mass-damper system

$$v_{mH} = Ae^{-bt/m}$$

(2) Find the particular solution. This is the value of the variable as t approaches infinity and can be found from the circuit by shorting T-type storage and opening A-type storage. For the mass-damper the mass acts as an open circuit and therefore

$$v_m(\infty) = v(\infty) = 2 \text{ m/s}$$

(3) Find the initial condition of the variable. Immediately upon application of a step the T-type elements are open and A-type elements are short. The mass thus behaves initially as a short circuit and $v_m(0^+) = 0$.

(4) Add the homogeneous and particular solutions together to get the complete solution

$$v_m(t) = Ae^{-bt/m} + 2$$

and then evaluate A from the initial condition $v_m(0^+) = 0$. Thus:

$$v_m(t) = 2(1 - e^{-t/10}) \, u_s(t)$$

represents the complete solution as before.

In this example the utility and advantage of the physical approach is not obvious. When we deal with responses that depend on the derivatives of inputs the benefits of the physical approach will become clear.

Let us continue with another two element circuit. However, in this case the storage element and the output variable associated with it are of different types (e.g., a capacitor and the current through it or an inductor and the voltage across it). As a result $a_1 = 0$ and Equation (5.1) may be written as:

$$\tau \frac{dy}{dt} + y = a_2 \frac{dg(t)}{dt} \qquad (5.14)$$

We call this a type 2 circuit. In this circuit the output variable, $y(t)$, can change suddenly so that in general $y(0^+) \neq 0$. If $g(t) = G u_s(t)$ then Equation (5.14) becomes:

$$\tau \frac{dy}{dt} + y = a_2 G u_i(t) \qquad (5.15)$$

Thus the application of a step input to the type 2 circuit leads to a differential equation with an impulse function on the right-hand side. We will first obtain a mathematical solution of Equation (5.15). Then in Example 5.2 which follows we will use the physical approach.

If we integrate Equation (5.15) with respect to time the result is:

$$\tau \frac{d}{dt} [\int_0^t y \ dt'] + [\int_0^t y \ dt'] = a_2 G u_s(t) \qquad (5.16)$$

Suppose we now define the new variable, $z(t)$, where

$$z(t) = \int_0^t y \ dt' \qquad (5.17)$$

When Equation (5.16) is expressed in terms of z we obtain:

$$\tau \frac{dz}{dt} + z = a_2 G u_s(t) \qquad (5.18)$$

Furthermore we recognize that from Equation (5.17), $z(0^+) = 0$ since the upper and lower limits of integration have the same value. Equation (5.18) is now precisely the

same form as Equation (5.3). Therefore the solution for
z is the same as that given in Equation (5.13) for y,
namely:

$$z(t) = a_2 G(1 - e^{-t/\tau})u_s(t) \qquad (5.19)$$

This is a solution for z(t) and we seek a solution for
y(t). The relation between the functions is given in
Equation (5.17). Therefore from Equations (5.17) and
(5.19) we find that:

$$y(t) = \frac{dz(t)}{dt} = \frac{a_2 G}{\tau} e^{-t/\tau} u_s(t) + a_2 G[1 - e^{-t/\tau}]u_i(t)$$

$$(5.20)$$

The last term on the right-hand side of Equation (5.20)
is identically equal to zero. This may be explained by
noting that any function g(t) multiplied by an impulse

$u_i(t - t_o)$ is:

$$g(t)u_i(t - t_o) = g(t_o)u_i(t - t_o)$$

Thus:

$$(1 - e^{-t/\tau})u_i(t) = u_i(t) - e^{-t/\tau} u_i(t)$$
$$= u_i(t) - (1)u_i(t) = 0$$

Thus the solution for the step input in a type 2 circuit
is:

$$y(t) = \frac{a_2 G}{\tau} e^{-t/\tau} u_s(t) \qquad (5.21)$$

Figure 5.4 shows the normalized step response described
by Equation (5.21). The output variable changes suddenly
at t = 0 and thereafter has an exponential decay. The
following example pertains to a type 2 circuit.

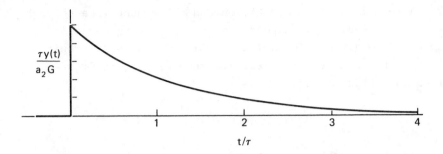

Figure 5.4 Normalized step response of type 2 circuit

Example 5.2

A pump supplies water to a drum through a long line as shown in Figure 5.5a. The pump is suddenly turned on and may be modelled as $p(t) = 2(10)^4 u_S(t)$, Pa. The line has a fluid resistance, $R = 500$ N·s/m^5 and the drum has a fluid capacitance, $C = 0.8(10)^{-2}$ m^5/N. Find an expression for the flow rate into the drum, $q(t)$.

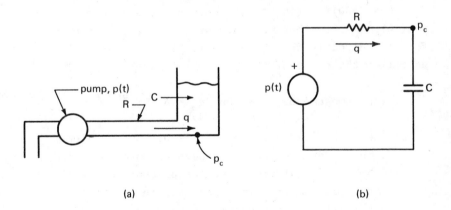

    (a)   (b)

Figure 5.5 Fluid circuit (example 5.2)

Figure 5.5b shows the circuit that represents the physical configuration of the problem. The drum is a fluid capacitance and therefore an A-type storage device.

The output variable, q(t), on the other hand is a through type variable. We expect, then, a governing equation of the form given in Equation (5.14) for a type 2 circuit. From the compatibility relation the circuit in Figure 5.5b yields:

$$[p(t) - p_c] + [p_c] = p(t)$$

From the elemental equations, $p(t) - p_c = Rq$ and $p_c = (1/C) \int q \, dt$ so that:

$$Rq + \frac{1}{C} \int q \, dt = p(t)$$

If we take the derivative of the above equation the result is:

$$RC \frac{dq}{dt} + q = C \frac{dp(t)}{dt}$$

This equation is in the form of Equation (5.14) with $\tau = RC = (500)(.8)(10)^{-2} = 4$ seconds and $a_2 = C = 0.8(10)^{-2}$ m^5/N. Thus we could write the expression for q(t) directly from Equation (5.21). However, let us follow the steps of the physical approach.

(1) The homogeneous solution, $q_H = A \, e^{-t/RC}$

(2) The particular solution $q(\infty) = 0$. After a long time the drum capacitance becomes an open circuit. From a physical viewpoint the pressure at the bottom of the drum increases as the drum fills. Ultimately, a pressure is reached that is sufficiently high to prevent further flow passing from pump to drum.

(3) The initial condition. From the circuit (Figure 5.5b) we recognize that the drum acts as a short circuit for sudden changes. Thus instantaneously

the circuit flow depends only on the resistance and pump pressure.

$$q(0^+) = \frac{P(0^+)}{R} = \frac{2(10)^4}{500} = 40 \text{ m}^3/\text{s}$$

(4) The complete solution is the sum of the homogeneous and particular solutions

$$q(t) = A\ e^{-t/RC} + 0$$

and A may be evaluated from the initial condition so that:

$$q(t) = 40\ e^{-t/4}\ u_S(t)$$

This is the same result that would be obtained from Equation (5.21).

Let us return now to the general differential equation for a first-order circuit given in Equation (5.1).

$$\tau \frac{dy}{dt} + y = a_1\ g(t) + a_2\ \frac{dg(t)}{dt} \qquad (5.1)$$

We may decompose Equation (5.1) into equations that have already been solved.

$$\tau \frac{dy_1}{dt} + y_1 = a_1\ g(t) \qquad (5.22a)$$

$$\tau \frac{dy_2}{dt} + y_2 = a_2\ \frac{dg(t)}{dt} \qquad (5.22b)$$

$$y = y_1 + y_2 \qquad (5.22c)$$

The step response solution for Equations (5.22a) and (5.22b) is given in Equations (5.13) and (5.21) respectively as:

$$y_1(t) = a_1\ G[1 - e^{-t/\tau}]u_S(t) \qquad (5.23a)$$

$$y_2(t) = \frac{a_2 G}{\tau} e^{-t/\tau} u_s(t) \qquad (5.23b)$$

The complete step response solution for the general first order circuit is, therefore, the sum of Equations (5.23a) and (5.23b).

$$y(t) = [a_1 G + (\frac{a_2}{\tau} - a_1)G \, e^{-t/\tau}]u_s(t) \qquad (5.24)$$

We will make use of another example to show that the physical approach produces the same result.

Example 5.3

Figure 5.6 shows an electrical circuit which has a voltage source $e(t) = E \, u_s(t)$. Find the current $i(t)$ through resistor, R_1.

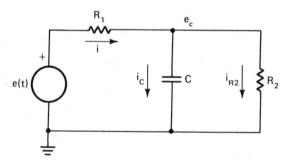

Figure 5.6 Electrical circuit (example 5.3)

From the compatibility relation (KVL) we may write

$$[e(t) - e_c] + e_c = e(t)$$

This expression may be applied for the loop 1 including R_1 and C and also for the loop 2 including R_1 and R_2. Substitution of the elemental equations then yields:

$$\text{For loop 1} \quad \rightarrow \quad R_1 \, i + \frac{1}{C} \int i_c \, dt = e(t)$$

$$\text{For loop 2} \quad \rightarrow \quad R_1 \, i + R_2 \, i_{R_2} = e(t)$$

If the equation from loop 1 is differentiated with respect to time and multiplied by C we may rewrite the above as:

$$\text{For loop 1} \quad \rightarrow \quad R_1 C \frac{di}{dt} + i_c = C \frac{de(t)}{dt}$$

$$\text{For loop 2} \quad \rightarrow \quad \frac{R_1}{R_2} i + i_{R_2} = \frac{e(t)}{R_2}$$

If we add these equations and note that $i = i_c + i_{R_2}$ we obtain the differential equation for this circuit as:

$$\frac{R_1 R_2}{R_1 + R_2} C \frac{di}{dt} + i = \frac{e(t)}{R_1 + R_2} + \frac{R_2 C}{R_1 + R_2} \frac{de(t)}{dt}$$

This is in the form of Equation (5.1) with $\tau = R_1 R_2 C/(R_1+R_2)$, $a_1 = 1/(R_1+R_2)$ and $a_2 = R_2 C/(R_1+R_2)$. The solution can be found directly from Equation (5.24) with $G = E$. In the physical approach, however, we merely determine i_H, $i(0^+)$ and $i(\infty)$. The instantaneous value of the current is $i(0^+) = E/R_1$. This is a consequence of the short circuit behavior of the capacitance at $t = 0^+$. After a long time the capacitance becomes an open circuit so that $i(\infty) = E/(R_1+R_2)$. The homogeneous solution of all first-order circuits is of the same form, specifically $i_H = A e^{-t/\tau}$. Thus, the current $i(t)$ may be expressed as:

$$i(t) = i_H + i(\infty)$$

$$= A e^{-t/\tau} + E/(R_1+R_2)$$

Now from the initial condition, $i(0^+) = E/R_1$ we may evaluate A as:

$$i(0^+) = A + \frac{E}{R_1+R_2} = \frac{E}{R_1}$$

and therefore $A = R_2 E/R_1(R_1+R_2)$. Thus the complete step response of the circuit in Equation (5.6) is:

$$i(t) = \left[\frac{E}{R_1 + R_2} + \frac{R_2 E}{R_1 (R_1 + R_2)} e^{-t/\tau} \right] u_s(t)$$

$$= \frac{E}{R_1 + R_2} \left[1 + \frac{R_2}{R_1} e^{-t/\tau} \right] u_s(t)$$

Figure 5.7 shows the normalized step response for this circuit. The magnitude of the current depends on the ratio of R_2/R_1.

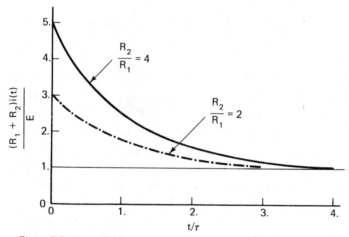

Figure 5.7   Normalized step response of electrical circuit (example 5.3)

5.2   RAMP, PARABOLIC AND IMPULSE RESPONSES

In the previous section we have developed the response of a first-order circuit to a step input.   Now the response, $y(t)$, to any general input function, $g(t)$ can also be obtained by direct mathematical solution of Equation (5.1). However, when the input function is related to the step function by integration or differentiation a simpler method is possible.   In these cases we may make use of a property of linear differential equations to determine the responses without solving the equations.   To illustrate the rationale behind this procedure suppose we designate the unit step

response for the general first-order circuit as y_s. Then
we may rewrite Equation (5.1) as:

$$\tau \frac{dy_s}{dt} + y_s = a_1 \, u_s(t) + a_2 \, u_i(t) \qquad (5.25)$$

If we subject the same circuit to a ramp input and call
the response, y_r, then Equation (5.1) becomes:

$$\tau \frac{dy_r}{dt} + y_r = a_1 \, u_r(t) + a_2 \, u_s(t) \qquad (5.26)$$

Now if Equation (5.25) is integrated with respect to
time the result is:

$$\tau \frac{d[\int_o^t y_s \, dt']}{dt} + [\int_o^t y_s \, dt'] = a_1 \, u_r(t) + a_2 \, u_s(t) \qquad (5.27)$$

Let us compare Equations (5.26) and (5.27). The right-
hand sides are identical. Thus the left-hand sides must
also be equal to each other. This condition is satisfied
only if

$$y_r = \int_o^t y_s \, dt' \qquad (5.28)$$

The result given in Equation (5.28) shows that the ramp
response of a linear circuit is equal to the integral of
the circuit step response. We may generalize this concept.
Figure 5.8a shows a block diagram in which the input $u_s(t)$
produces the output response $y_s(t)$. Figure 5.8b shows the
circuit with ramp input $u_r(t)$ and output response $y_r(t)$.
Since the input in Figure 5.8b is the integral of the input
in Figure 5.8a, the output of Figure 5.8b is the integral
of the output of Figure 5.8a. This is merely a restatement
of what we have proved in Equation (5.28). In a similar
way the parabolic and impulse responses may be related to
the step response. Figure 5.8c shows the circuit with an
input $u_p(t)$. The output response in this case, y_p, is
the double integral of the step response or the single
integral of the ramp response. The circuit with impulse

function input, $u_i(t)$, is shown in Figure 5.8d. The input
is the derivative of a step input. Therefore, the corresponding impulse response, $y_i(t)$, is the derivative of
the step response.

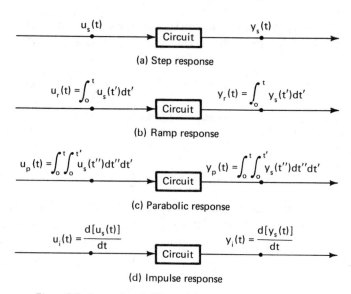

Figure 5.8   Integral and differential property of linear circuits

We may now apply this integral and differential property
of linear differential equations to find the ramp, parabolic and impulse responses of the general first-order
circuit. Equation (5.24) gives the step response as:

$$y_s(t) = [a_1 + (\frac{a_2}{\tau} - a_1) e^{-t/\tau}]G \ u_s(t) \quad (5.29)$$

Thus, the general ramp response is:

$$y_r(t) = \int_o^t y_s(t')dt'$$

$$= [a_1 t + (a_2 - a_1\tau)(1 - e^{-t/\tau})]G \ u_s(t) \quad (5.30)$$

The parabolic response is:

$$y_p(t) = \int_o^t y_r(t')dt' = \int_o^t \int_o^{t'} y_s(t'')dt''dt'$$

$$= \left[\frac{a_1 t^2}{2} + (a_2 - a_1\tau)(t + \tau - \tau e^{-t/\tau}) \right] G\, u_s(t) \quad (5.31)$$

and the impulse response is:

$$y_i(t) = \frac{d(y_s(t))}{dt}$$

$$= \frac{a_2 G}{\tau} u_i(t) - \frac{1}{\tau} (\frac{a_2}{\tau} - a_1)G\, e^{-t/\tau}\, u_s(t) \quad (5.32)$$

The solutions given in Equations (5.29), (5.30), (5.31) and (5.32) are valid for the step, ramp, parabolic and impulse responses of all first-order circuits with initial rest conditions. In the following examples, however, we choose to extend the physical approach rather than to rely on these general formulas. When the input is not a step function we add a fifth step to the physical approach. The first four steps remain the same and the complete step response is obtained. In step 5 we perform a mathematical operation on the step response to get the desired response.

Example 5.4

A mechanical spring and damper are connected together as shown in Figure 5.9a. The spring constant $k = 25$ N/m and the damping $b = 75$ N$\cdot$s/m. Find the velocity of the connecting point if the applied force, $f(t)$ is:

(a) $50u_r(t)$, N   and   (b) $50u_i(t)$, N

The analogous circuit for the mechanical arrangement is shown in Figure 5.9b. To find the differential equation we apply the continuity equation at the node, v. Thus:

$$f_b + f_k = f(t)$$

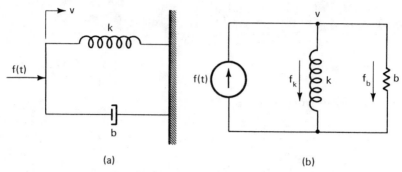

Figure 5.9 Mechanical circuit (example 5.4)

Now by substituting the elemental equations into the above we obtain:

$$bv + k \int v \, dt = f(t)$$

This equation can be placed in the standard form by differentiating and dividing by k.

$$\frac{b}{k} \frac{dv}{dt} + v = \frac{1}{k} \frac{df(t)}{dt}$$

We will first find the response of the system to the step force, $f(t) = 50u_s(t)$, N.

1) The homogeneous solution

$$v_H = A \, e^{-kt/b} = A \, e^{-t/3}$$

can be written by inspection.

2) The particular solution

A long time after a sudden change is made the spring behaves like a short circuit. Thus the velocity is grounded and we may write:

$$v(\infty) = v_P = 0$$

3) The initial condition

The short-time performance of a spring is similar to an open circuit. For a sudden change, the damper takes all the force instantaneously. We may, therefore, specialize the elemental equation of the damper at $t = 0^+$ as:

$$f_b(0^+) = f(0^+) = bv(0^+)$$

or

$$v(0^+) = 50/75 = 2/3$$

4) The complete step response, v_s is:

$$v_s(t) = A\ e^{-t/3} + 0$$

When the initial condition is used, the value of A is determined as 2/3 and the complete solution is:

$$v_s(t) = (2/3)e^{-t/3}\ u_s(t),\ m/s$$

5) To find the ramp response, v_r, and the impulse response, v_i, recall that:

$$v_r = \int_o^t v_s(t')dt' \quad \text{and} \quad v_i = \frac{dv_s}{dt}$$

so that

$$v_r(t) = 2(1 - e^{-t/3})u_s(t),\ m/s$$

and

$$v_i(t) = (2/3)u_i(t) - (2/9)e^{-t/3}\ u_s(t),\ m/s$$

These results agree with those that would be obtained from Equations (5.30) and (5.32) with:

$$\tau = 3, \quad a_1 = 0, \quad a_2 = 1/25 \quad \text{and} \quad G = 50$$

The responses are shown graphically in Figures 5.10a and b. For a ramp force input the system approaches a constant velocity. When the force input is an impulse the system moves instantaneously to some position where the spring is depressed. Thereafter, the system returns slowly to its static equilibrium position.

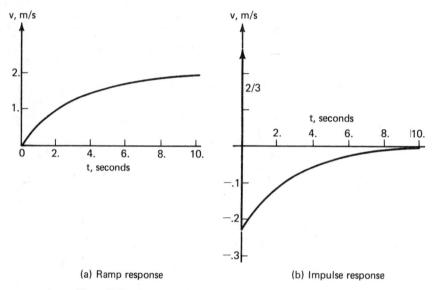

(a) Ramp response   (b) Impulse response

Figure 5.10   Responses for mechanical circuit in example 5.4

Example 5.5

Figure 5.11 shows an RL series electric circuit. If R = 10 ohms and L = 1 H, find the current, i(t), for voltage inputs:

(a) $e(t) = 5u_r(t)$, volts   and   (b) $e(t) = 20u_p(t)$, volts

To find the governing differential equation we apply the compatibility condition around the circuit.

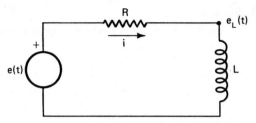

Figure 5.11 Electric circuit (example 5.5)

$$e_L + (e - e_L) = e$$

Substitution of the elemental relations leads directly to:

$$L \frac{di}{dt} + Ri = e(t)$$

or in standard form:

$$\frac{L}{R} \frac{di}{dt} + i = \frac{1}{R} e(t)$$

We follow the usual procedure now and find the solution of the circuit when $e(t) = u_s(t)$, volts.

1) The homogeneous solution

$$i_H = A\, e^{-Rt/L} = A\, e^{-10t}$$

is found directly since this is always the form for a first-order circuit.

2) The particular solution
After a long time following the step input the inductance behaves as a short circuit. Thus:

$$i_P = i(\infty) = e(\infty)/R = 0.1$$

3) The initial condition
   Initially the inductor behaves as an open circuit.
Therefore:

$$i(0^+) = 0$$

4) The complete step response is:

$$i_s(t) = A\, e^{-10t} + 0.1$$

and with the initial condition used A = -0.1 so that we
may write:

$$i_s(t) = 0.1(1 - e^{-10t})u_s(t), \text{ A}$$

5) (a) The response to $5u_r(t)$ is $i_r(t)$ and may be ex-
       pressed as:

$$i_r(t) = 5 \int_0^t i_s(t')dt'$$

   and when the integration operation on the step
   response is performed

$$i_r(t) = .5(t - 0.1 + 0.1\, e^{-10t})u_s(t)$$

   This response is shown plotted in Figure 5.12a.
   Note that the asymptote in this form of ramp
   response always intersects the time axis at $t = \tau$.

   (b) The response to $20u_p(t)$ is $i_p(t)$ and is defined
       as:

$$i_p(t) = 20 \int_0^t \int_0^{t'} i_s(t'')dt''dt'$$

   If the double integration is performed the result
   is:

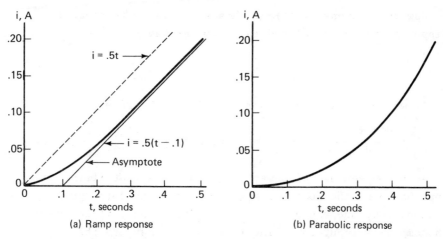

Figure 5.12 Responses for electric circuit in example 5.5

$$i_p(t) = 2[t^2/2 - 0.1\,t + 0.01 - 0.01\,e^{-10t}]u_s(t)$$

The parabolic response is shown plotted in
Figure 5.12b.
The response may be checked from Equations (5.30)
and (5.32).

5.3 RESPONSE TO COMPOSITE FUNCTIONS

In the previous section we observed one of the important
properties of a linear system. Specifically if an input,
$g(t)$, produces a response, $y(t)$ then other inputs that
are related mathematically to the original input (i.e.
$dg(t)/dt$ or $\int g(t)dt$) produce responses with the same
mathematical operation (i.e., $dy(t)/dt$ or $\int y(t)dt$).
Another important property of linear systems is superposition. Consider that the response $y_1(t)$ to an input
$g_1(t)$ for the general first-order circuit is described
by the differential equation:

$$\tau\,\frac{dy_1}{dt} + y_1 = a_1\,g_1 + a_1\,\frac{dg_1}{dt} \qquad (5.33)$$

Now the response $y_2(t)$ of this same system to another
input $g_2(t)$ is formulated in a similar way as:

$$\tau \frac{dy_2}{dt} + y_2 = a_2 \, g_2 + a_2 \frac{dg_2}{dt} \tag{5.34}$$

If we add Equations (5.33) and (5.34) the result is:

$$\tau \frac{d(y_1 + y_2)}{dt} + (y_1 + y_2) = a_1(g_1 + g_2) + a_2 \frac{d(g_1 + g_2)}{dt} \tag{5.35}$$

As a consequence of Equation (5.35) note that if the input had been the function $g_1(t) + g_2(t)$, the corresponding
output response would be $y_1(t) + y_2(t)$. Thus, if we apply
a composite input to a linear circuit we may use the component portions of the input separately to obtain an output for each. The complete response to a composite input
is then the sum of the output component portions. In more
general terms we may state for the above inputs and outputs that an input of $A_1 g_1(t) + A_2 g_2(t)$ will produce an
output of $A_1 y_1(t) + A_2 y_2(t)$ where A_1 and A_2 are constants.
   In addition to the above properties we may also time
shift the response of linear circuits. For example, if
input $g(t)$ causes output $y(t)$ then input $g(t-T)$ will cause
output $y(t-T)$. The shifting feature of linear circuits
with zero initial conditions is one we will often use in
conjunction with composite input functions. The following
examples will clarify the superposition and shifting
properties.

Example 5.6

   Figure 5.13a shows a first-order mass-damper system
with the mass, m = 100 kg and the damper, b = 20 N·s/m.
A force, f(t), of the form shown in Figure 5.13b is
applied to the system.
   Determine the velocity of the mass at the end of 1
second, v(1), and at the end of 3 seconds, v(3).
   The analogous circuit for this mechanical system is
presented in Figure 5.14. Remember that the mass always

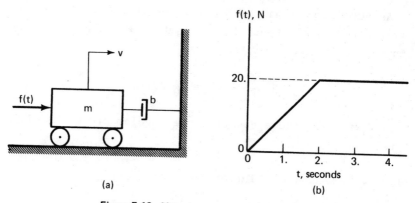

(a)

(b)

Figure 5.13 Mass-damper system (example 5.6)

appears as a capacitance to ground. The continuity
condition (KCLA) at either node, yields:

$$f_m + f_b = f(t)$$

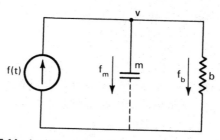

Figure 5.14 Analogous circuit for mass-damper system (example 5.6)

From the elemental relations we know immediately that
$f_m = 100 \, dv/dt$ and $f_b = 20 \, v$. The input function, $f(t)$,
may be separated into the sum of two ramp functions,

$$f(t) = 10u_r(t) - 10u_r(t - 2)$$

Thus the system equation is:

$$5 \frac{dv}{dt} + v = \frac{f(t)}{20}$$

As before, we begin by finding the response of the
system to a step input, $f(t) = u_s(t)$. For the step
input the system equation is:

$$5 \frac{dv_s}{dt} + v_s = \frac{1}{20} u_s(t)$$

Now with the above equation the regular procedure will
yield $v_s(t)$.

1) The homogeneous solution

$$v_H = A e^{-.2t}$$

2) The particular solution
  A capacitor does not pass steady current. It behaves
as an open circuit when the voltage across it is steady.
The analogous mechanical component, mass, takes no force
a long time after a sudden change. In this system all
the force supplied after a long time is used to overcome
damping. As a result $v_p = f(\infty)/b = 1/20$.

3) The initial condition
  A mass initially at rest cannot suddenly have an
initial velocity with a finite driving force and
$v_s(0^+) = 0$.

4) The complete step response is the sum of the homogeneous and particular solutions.

$$v_s(t) = A e^{-.2t} + \frac{1}{20}$$

and from the initial condition $A = -1/20$, so that

$$v_s(t) = \frac{1}{20} (1 - e^{-.2t})u_s(t), \ m/s$$

5) (a) The response to $10u_r(t)$ is $v_1(t)$ where

$$v_1(t) = 10 \int v_s(t')dt'$$

$$v_1(t) = 0.5[t + 5e^{-.2t} - 5]u_s(t)$$

(b) The response to $-10u_r(t - 2)$ is $-v_1(t - 2)$ by the shifting property of linear equations

$$- v_1(t - 2) = - 0.5 [(t - 2) + 5e^{-.2(t-2)} - 5]u_s(t - 2)$$

This is the same function as $v(t)$ with all the t arguments replaced by $(t - 2)$. The complete response to the given function, $f(t)$, is therefore the superposition of parts (a) and (b) or

$$v(t) = 0.5[t + 5e^{-.2t} - 5]u_s(t) - 0.5[(t - 2)$$

$$+ 5e^{-.2t(t-2)} - 5]u_s(t - 2)$$

6) Evaluation of the response at the desired times of 1 and 3 seconds. From the complete response we obtain:

$$v(1) = 0.5[1 + 5e^{-.2} - 5] - 0 = 0.047 \text{ m/s}$$

Note that the second term in the response does not contribute to $v(1)$. The output response can never come before the input which produced it.

$$v(3) = 0.5[3 + 5e^{-.6} - 5] - 0.5[1 + 5e^{-.2} - 5]$$

$$= 0.325 \text{ m/s}$$

Example 5.7

In the series RC electric circuit (Figure 5.15a), the value of R is 100 ohms and the value of C is 0.1 F. Find the voltage drop across the resistor $e_R(t)$ when the voltage source $e(t)$ is the rectangular pulse shown in Figure 5.15b. Compare this response to the response when $e(t) = 20u_i(t)$, volts.

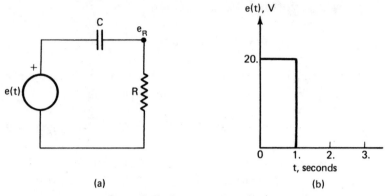

Figure 5.15   Electrical circuit (example 5.7)

From the continuity condition at the node, e_R, and the elemental equations the differential equation for this circuit is:

$$10 \frac{de_R}{dt} + e_R = \frac{de(t)}{dt}$$

The rectangular pulse shown in Figure 5.15b may be described in terms of singularity functions as:

$$e(t) = 20u_s(t) - 20u_s(t - 1)$$

We start by finding the response of the circuit to the unit step function ($e(t) = u_s(t)$).

1) The homogeneous solution, $e_H = A\, e^{-t/10}$

2) The particular solution, $e_P = 0$

3) The initial condition, $e_R(0^+) = e(0^+) = 1$

4) The complete solution, $e_R(t) = e^{-t/10}\, u_s(t)$

This solution is the response to a unit step function. The response to the rectangular pulse function given is, therefore, by superposition and shifting:

$$e_{RP}(t) = 20e^{-t/10} u_s(t) - 20e^{-(t-1)/10} u_s(t - 1)$$

The response to the impulse function is:

$$e_{Ri}(t) = 20 \frac{de_R}{dt} = 20u_i(t) - 2e^{-t/10} u_s(t)$$

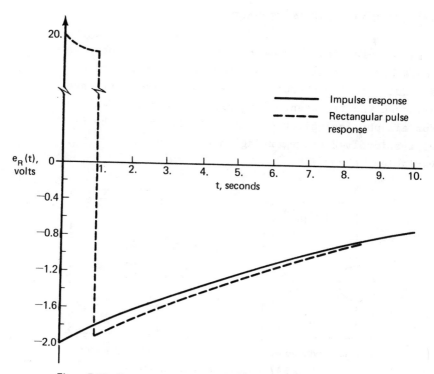

Figure 5.16 Comparison of impulse and rectangular pulse response (example 5.7)

The rectangular pulse and impulse responses are shown plotted in Figure 5.16. Both responses consist of two terms. The impulse response (solid line) has a positive impulse at $t = 0$ and a negative decaying exponential. The rectangular pulse response (dashed line) represents the difference between an exponential function and an exponential function delayed. At $t = 1$ the rectangular pulse response changes discontinuously. Note that as time increases the two responses approach each other.

At t = 10 seconds the difference between the responses
is 0.038 volts. We may state, in general, that a
rectangular pulse of short duration compared to the
circuit time constant produces a response similar to
an impulse of the same strength.

5.4 RESPONSE WITH INITIAL STORAGE

In the circuits previously considered the storage elements had no initial energy storage. We now turn our
attention to first-order circuits with energy storage.
Although the treatment of initial storage is identical
for all physical circuits, we will illustrate the principles involved by referring to the electrical signal
variables, e and i.

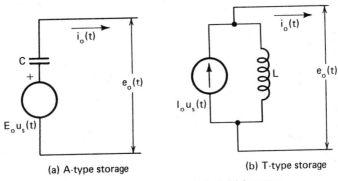

(a) A-type storage   (b) T-type storage

Figure 5.17 Circuit models for initial storage

Figure 5.17 reviews the circuit models for initial
storage that were developed in Chapter IV. For A-type
storage a potential source, $E_0 \, u_s(t)$, is placed in series
with the capacitive component. For T-type storage a flow
source, $I_0 \, u_s(t)$, is placed in parallel with the inductive
component. E_0 represents the magnitude of the initial
voltage across the capacitor and I_0 represents the magnitude of the initial current through the inductor. When
a component has initial storage we replace the component

by one of the circuit models for initial storage shown in
Figure 5.17.

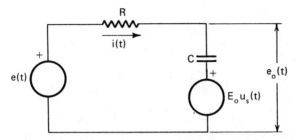

Figure 5.18 Series RC circuit with initial storage

Figure 5.18 shows the equivalent circuit for a series
RC with initial storage. Note that the storage element
(the capacitor) is replaced by a storage element and a
source in series. Suppose we wish to determine the
current response i(t), and the voltage response across
the capacitor, $e_0(t)$, for a source input, e(t). To obtain
these responses we make use of the principle of superposition of sources. The circuit is treated as if it
had two separate sources:
  1) the input source, e(t) and
  2) the storage source, $E_0 u_s(t)$

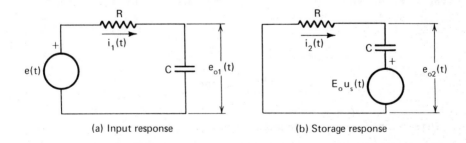

(a) Input response   (b) Storage response

Figure 5.19 Separation of sources in series RC circuit with storage

The effect of each source is then added to form the complete response. Figure 5.19a shows the input source acting

alone. In this diagram the A-storage source is zero or a
short circuit. The resulting circuit is then the circuit
without initial storage that we have already analyzed.
The circuit equations without storage is:

$$RC \frac{de_{o_1}}{dt} + e_{o_1} = e(t) \tag{5.36a}$$

$$RC \frac{di_1}{dt} + i_1 = C \frac{de(t)}{dt} \tag{5.36b}$$

Figure 5.19b shows the storage source acting alone. The
input voltage source is zero, in this case, and therefore
it has been replaced by a short circuit. The circuit
equations without an input source are:

$$RC \frac{de_{o_2}}{dt} + e_{o_2} = RC \frac{d}{dt} [E_o\, u_s(t)] = RC\, E_o\, u_i(t) \tag{5.37a}$$

$$RC \frac{di_2}{dt} + i_2 = - C \frac{d}{dt} [E_o\, u_s(t)] = - C\, E_o\, u_i(t) \tag{5.37b}$$

Now by superposition of sources the complete response
for the voltage, $e_o = e_{o_1} + e_{o_2}$ and for the current
$i = i_1 + i_2$. The input response portion of the complete
response has been dealt with in the previous sections and
remains unchanged. It will, of course, depend on the
input function applied. The storage portion of the
response, on the other hand, can be obtained by solving
Equation (5.37). The result is:

$$e_{o_2} = E_o\, e^{-t/RC}\, u_s(t) \tag{5.38a}$$

$$i_2 = - \left(\frac{E_o}{R}\right) e^{-t/RC}\, u_s(t) \tag{5.38b}$$

This portion of the response does not depend on the
input function and is, therefore, the same for any input
function. The storage response is always in the characteristic form $A\, e^{-t/\tau}$ where A is the initial value of the
variable due to storage. Let us demonstrate that this
result holds also for T-type storage.

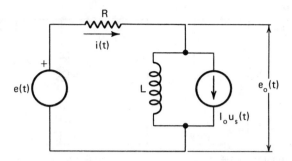

Figure 5.20 Series RL circuit with initial storage

Figure 5.20 shows a series RL circuit with initial storage. The inductor has been replaced by an inductor and flow source in parallel. To find the complete responses, e_o and i, we may treat each source separately and then superimpose the results. Figure 5.21a shows the circuit with the flow storage source equal to zero or open circuited. This circuit is an RL series circuit without storage. The responses due to the input source depend on solutions to the equations:

$$\frac{L}{R}\frac{de_{o_1}}{dt} + e_{o_1} = \frac{L}{R}\frac{de(t)}{dt} \qquad (5.39a)$$

$$\frac{L}{R}\frac{di_1}{dt} + i_1 = \frac{1}{R}e(t) \qquad (5.39b)$$

This portion of the response we have found in the previous sections. It depends on the type of input function, $e(t)$ that is applied.

Figure 5.21b shows the series RL circuit with the input voltage source set equal to zero. The circuit differential equations for this portion of the response are:

$$\frac{L}{R}\frac{de_{o_2}}{dt} + e_{o_2} = L\frac{d}{dt}[I_o\,u_s(t)] = L\,I_o\,u_i(t) \qquad (5.40a)$$

$$\frac{L}{R}\frac{di_2}{dt} + i_2 = -\frac{L}{R}\frac{d}{dt}[I_o\,u_s(t)] = -\frac{L}{R}\,I_o\,u_i(t) \qquad (5.40b)$$

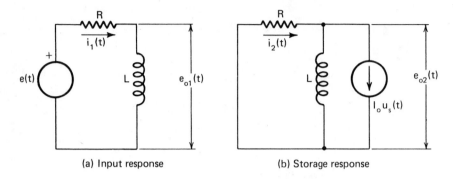

(a) Input response   (b) Storage response

Figure 5.21   Separation of sources in series RL circuit with storage

The storage portion of the response for the RL circuit can be determined by solving Equation (5.40) and is:

$$e_{o_2} = RI_o \, e^{-Rt/L} \, u_s(t) \qquad\qquad (5.41a)$$

$$i_2 = -I_o \, e^{-Rt/L} \, u_s(t) \qquad\qquad (5.41b)$$

Thus as in the RC circuit the response due to storage in the RL circuit has the characteristic form $A \, e^{-t/\tau}$ where A is the initial value of the variable due to storage. When the output current response is desired care must be exercised to make sure the sign of the storage contribution is correctly handled.   The following examples will clarify the solutions of first-order circuits with storage.

Example 5.8

A leaky water tank is being filled from an overhead pipe (Figure 5.22a).   The tank has a fluid capacitance $C = 0.01 \, m^5/N$ and the leak may be modelled as a linear fluid resistance $R = 1500 \, N \cdot s/m^5$.   The flow into the tank, $q(t)$, has the function shown in Figure 5.22b.   At the instant the flow begins $(t = 0)$ the height of the water in the tank is 10 m.   Find:

(a) the pressure at the bottom of the tank, p(t)

(b) the height of the water after 20 seconds.

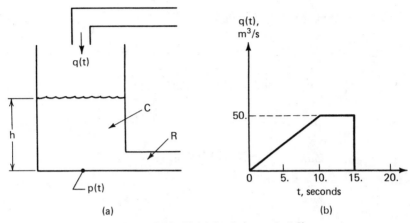

Figure 5.22   Fluid circuit (example 5.8)

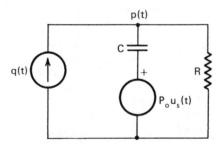

Figure 5.23   Analogous circuit for example 5.8

The analogous circuit is shown in Figure 5.23.   As previously indicated we obtain separately the responses due to the flow and storage sources.

Step Response due to q(t) = $u_S(t)$ - no storage.

1) The homogeneous solution, $p_H = A\ e^{-t/RC}$

2) The particular solution, $p_p = R(1)$

3) The initial condition - no storage, $p(0^+) = 0$

4) Unit step response, $p_s(t) = R(1 - e^{-t/RC})$

<u>Response</u> due to $q(t) = 5u_r(t) - 5u_r(t - 10) - 50u_s(t - 15)$

5) $p_1(t) = 5R[t + RC\, e^{-t/RC} - RC]u_s(t)$

$- 5R[(t - 10) + RC\, e^{-(t-10)/RC} - RC]u_s(t - 10)$

$- 50R[1 - e^{-(t-15)/RC}]u_s(t - 15)$

since

$R = 1\,500\ N{\cdot}s/m^5, \quad C = 0.01\ m^5/N$

$p_1(t) = 7\,500[t + 15e^{-t/15} - 15]u_s(t)$

$- 7\,500[(t - 10) + 15e^{-(t-10)/15} - 15]u_s(t - 10)$

$- 75\,000[1 - e^{-(t-15)/15}]u_s(t - 15)$

<u>Response</u> due to storage - no input.

6) $p_2(t) = \rho g h_o e^{-t/RC} = (1000)(9.81)(10)e^{-t/15}\, u_s(t)$

$= 98\,100e^{-t/15}\, u_s(t)$, Pa.

<u>Complete Response</u>

7) $p(t) = p_1(t) + p_2(t)$

$p(t) = 7\,500[t + 28.1e^{-t/15} - 15]u_s(t)$

$- 7\,500[(t - 10) + 15e^{-(t-10)/15} - 15]u_s(t - 10)$

$- 75\,000[1 - e^{-(t-15)/15}]u_s(t - 15)$

This is the solution to part (a).

$$p(20) = 7\ 500[5 + 7.4] - 7\ 500[-5 + 7.701]$$

$$- 75\ 000[1 - .717] = 51\ 475$$

$$h(20) = \frac{51\ 475}{9\ 810} = 5.25\ m$$

This is the solution to part (b).

Example 5.9

The mechanical arrangement shown in Figure 5.24a consists of a spring and damper. The spring has a constant of $k = 20$ N/m and is initially stretched a distance of 0.5 m. The damper has a value, $b = 10$ N·s/m. A velocity input $v(t) = u_r(t)$ is applied to the free end of the stretched spring. Find the reaction force that the mechanical circuit exerts on the wall ($f(t)$).

Figure 5.24b shows the analogous circuit diagram. From the initial stretch of the spring $F_O = kx = (20)(.5) = 10$ N.

<u>Response</u> to a unit step input - no storage.

1) The homogeneous solution $f_H = A\ e^{-kt/b}$

2) The particular solution $f_P = b(1)$

3) The initial condition $f(0) = 0$

4) The complete solution $f_S(t) = b(\ 1 - e^{-kt/b})u_S(t)$

<u>Response</u> to ramp input - no storage

5) $f_1(t) = b[t + \frac{b}{k}\ e^{-kt/b} - \frac{b}{k}]\ u_S(t)$

<u>Storage Response</u> - no input.

6) $f_2(t) = -\ F_O\ e^{-kt/b}\ u_S(t)$

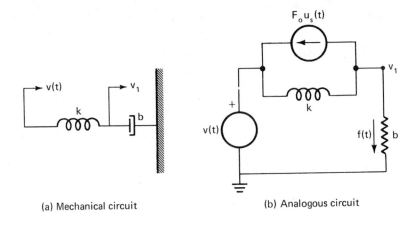

(a) Mechanical circuit   (b) Analogous circuit

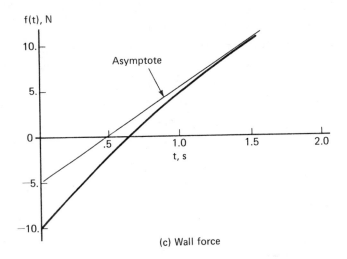

(c) Wall force

Figure 5.24 Circuit for example 5.9

Superposition of Responses

7) $f(t) = f_1(t) + f_2(t)$

$$= 10[t + (1/2)e^{-2t} - (1/2)]u_s(t) - 10e^{-2t} u_s(t)$$

$$f(t) = [10t - 5e^{-2t} - 5]u_s(t), \text{ N}$$

Initially, the force on the wall acts to the left.
When t = 0.64 there is no force on the wall and thereafter (t > 0.64) the force acts to the right.   The complete response is shown in Figure 5.24c.

Problems

5.1 In a carnival game contestants try to hit a lever
    hard enough to cause a mass to move up a pole and
    strike a bell.   The minimum force required for the
    mass to reach the bell is $70u_i(t)$, N.   If the mass
    is 3.0 kg and we assume that the friction between
    mass and pole may be modelled as a linear damper of
    5.0 N·s/m, find:
    (a) The time for the mass to reach the bell when
        the minimum impulsive force is applied.
    (b) The height of the bell, h.

5.2 An open water tank can be filled in two different
    ways: from the top and from the bottom.   In both
    schemes the source is a large reservoir of constant
    water level 8.16 m. The resistance R = 200 000 N·s/m^5,
    the capacitance C = 0.001 m^5/N and the density of
    water is 1 000 kg/m^3.   Find, for each method of fill-
    ing, the time for the water to reach a height of
    4.0 m in an initially empty tank.

5.3 The mass-damper system shown in Figure P5.3 is
    initially at rest.   If b_1 = 60 N·s/m, b_2 = 120 N·s/m
    and m = 175 kg, find:
    (a) The velocity of the mass at t = 0.25 s when
        $v(t) = 8u_s(t)$ m/s.
    (b) The time required for the mass to reach the
        velocity attained in part 'a', if the input
        velocity function v(t) is replaced by an input
        force function $f(t) = 100u_s(t)$, N.
    (c) The effect of a change in b_1 on the time required
        in part 'b'.

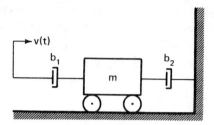

Figure P 5.3

5.4 A first-order mechanical system consists of two components in parallel connected at one end to a fixed
support. If the input force to the free end is
$600u_s(t)$, N and the resulting output velocity is
shown in Figure P5.4, identify the components and
find their values.

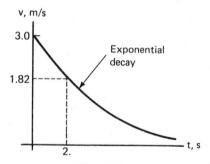

Figure P 5.4

5.5 The process of making wine involves 'slow cooking'
at 35°C above room temperature for several months.
The wine has a density of 1 000 kg/m³ and a specific
heat of 4 180 W·s/kg°C. To cook the wine a cubic
shaped container, 0.5 m on each side, is fabricated
with cork board 2 cm thick. The thermal conductivity of cork is 0.05 W/m°C. An electric light
bulb within the container acts as the heater. Find:
(a) The power required for the light bulb.
(b) The time constant of the system.

5.6 A step input $e(t) = 5u_s(t)$, V is applied to the electric circuit shown in Figure P5.6. If $R_1 = 1\ 000$ ohms and $R_2 = 4\ 000$ ohms what is the value of the capacitance so that $e_c(2) = 2.4$ V.

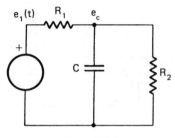

Figure P 5.6

5.7 A water tank is drained through a long pipeline. The tank has a capacitance $C = (10)^{-4}$ m^5/N and initially has 5.0 m height of water. If the pipe- line can be modelled as a pure resistance $R = 4.0(10)^5$ N·s/m^5, find:
   (a) the pressure at the bottom of the tank as a function of time after the tank valve is suddenly opened.
   (b) The height of the water in the tank after $t = 25$ s.

5.8 A golf ball lies on the green 10 m from the hole. The ball has a mass of 0.1 kg and the grass provides a damping of 0.038 N·s/m. The golfer applies a force with his putter that can be modelled as, $f(t) = 0.4u_i(t)$, N. If the ball moves in a straight line toward the hole, what is its velocity when it reaches the hole?

5.9 Consider the two RC circuits shown in Figure P5.9 connected to an ideal source. Find for each circuit:
   (a) $i(t)$ and $e_1(t)$ if the source is a voltage source $e(t) = 4.0u_s(t)$, V.
   (b) $i(t)$ and $e_1(t)$ if the source is a current source $i(t) = 20u_s(t)$, A.

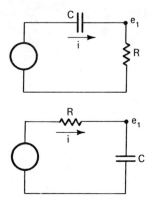

Figure P 5.9

5.10 The temperature inside a particular house is kept
     at 20°C. The heating system which maintains this
     temperature breaks down on a day when the outside
     temperature is a constant 5°C. After the break-
     down it takes 8 hours for the temperature inside
     to reduce to 15°C. The thermal resistance of the
     house may be represented by $R = 5(10)^{-8}°C \cdot hr/J$.
     Find:
     (a) The temperature in the house as a function of
         time.
     (b) The thermal capacitance of the house.

5.11 For the rotary mechanical system shown in Figure
     P5.11 angular velocity $\omega(t) = 5u_r(t) - 5u_r(t - 10)$,
     rad/s. If the damper $B = 20$ N·m·s and the moment
     of inertia of the mass is $J = 10$ kg·m², find:
     (a) $\omega_1(t)$.
     (b) The effect of a larger J on $\omega_1(t)$.

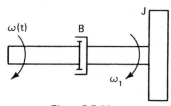

Figure P 5.11

5.12 A pump supplies water to the bottom of a large tank
  (C = 25 m^5/N) through a long line (R = 2.0 N·s/m^5).
  When there is initially no water in the tank the
  system requires that the pump be started and be capable
  of causing the tank level to reach 3.0 m in 50 s.
  (a) The chief engineer has selected a pump that
      develops a pressure 39.2(10)^3 u$_S$(t), Pa. Will
      this meet the requirements? What height is
      reached in 50 s?
  (b) An engineering student claims that the pump
      must deliver the pressure function, p(t) =
      A u$_S$(t) - (A - 39.2(10)^3)u$_S$(t - 10). What
      should be the value of A to satisfy the system
      requirements?

5.13 In a test to find the moment of inertia of the
  rotor of a generator the rotor is run up to a
  constant angular velocity and then the power
  source is turned off. The time for the rotor
  to reach a lower angular velocity is measured.
  Measurements show that the initial angular velocity
  is 500 r/min and it takes 3.5 minutes for the rotor
  to slow down to 400 r/min. If the retardation is
  due to friction which can be modelled as damping
  B = 110 N·m·s, what is the moment of inertia of
  the rotor?

5.14 A series RL electric circuit has R = 50 ohms and
  L = 2 H. If the voltage input given in Figure
  P5.14 is applied as a source, find the current and
  the voltage drop across the inductance as functions
  of time.

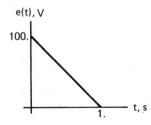

Figure P 5.14

5.15 The two horizontal force impulses, shown in Figure
     P5.15 are applied to a mass of 20 kg that is resting
     on a table. The time period, T, is the time required
     for the first impulse to cause the velocity of the
     mass to reach 0.0062 m/s. If the friction between
     mass and table may be modelled as viscous damping,
     b = 25 N·m/s, how far will the mass move from its
     initial position?

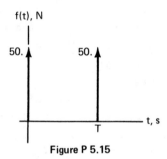

Figure P 5.15

5.16 The water in an electric water heater is initially
     at 20°C. An immersion heater is suddenly turned
     on and supplies 4 000$u_S(t)$, W. If the resistance
     of the insulation around the water tank is 0.02°C·s/J
     and the specific heat of water is 4 180 J/°C·kg
     find:
     (a) The mass of water that can be heated to 55°C
         in one hour.
     (b) The maximum temperature of the water after a
         long time.

5.17 In the rotational mechanical system shown in Figure
     P5.17 the flywheel is initially moving with an
     angular velocity of ω_0 rad/s clockwise. An external
     torque of T $u_S(t)$, N·m, is applied to the flywheel
     in a counterclockwise direction. Find the result-
     ing flywheel velocity in terms of J, B, T, ω_0 and t.

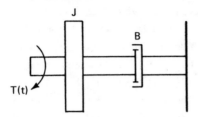

Figure P 5.17

5.18 A pipe carrying water is 10 m long and has a diameter of 2 cm. The pipe has a pump at one end and
is connected to atmosphere at the other end. The
pipe must be modelled with both resistance and
inertance. If the pump delivers P $u_s(t)$, Pa, find
the value of P that makes the Reynolds number in
the pipe 2 000 after 20 s.

5.19 In the mechanical system shown in Figure P5.19,
b = 260 N·s/m, k = 80 N/m, and the spring is initially compressed a distance of 0.25 m. If the
input force f(t) = $200u_s(t)$, N, find v(t) and x(t).

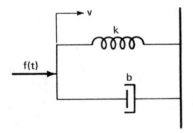

Figure P 5.19

5.20 An electric circuit contains a parallel combination
of R and L (R = 3.75 ohms and L = 1.5 H). The
circuit receives its input from a current source
$2u_r(t)$, A. However, the initial current in the
inductance is 1.0 A. Find the voltage drop across
the resistance as a function of time.

5.21 A pump delivering water to a tank through a resistive line (R_1 = 2 000 N·s/m^5) is suddenly turned
on.   Water flows out of the tank through another
resistance (R_2 = 8 000 N·s/m^5).   The capacitance
of the tank C = 9(10)$^{-3}$ m^5/N.   The output of the
pump may be modelled as P u_s(t), Pa.
   (a) What is the maximum value of P so that the
       water height in the tank never exceeds 5.0 m?
   (b) Repeat part 'a' if C = 20(10)$^{-3}$ m^5/N.

5.22 In the RC network shown in Figure P5.22, R_1 = 4 000 Ω,
R_2 = 6 000 Ω and C = 50 µF.   If e_i(t) = 200u_s(t),
find:
   (a) The current through the capacitor, i_c(t).
   (b) e_o(t).

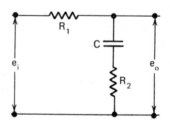

Figure P 5.22

5.23 To get from the first floor to ground in an emer-
gency, firemen often slide down a metal pole.   The
average fireman has a mass of 75 kg and requires
2 s to descend the distance, h, between first floor
and ground.   If the friction between the fireman
and pole can be modelled as damping b = 500 N·m/s,
find:
   (a) The velocity of the fireman when he reaches
       the ground.
   (b) The distance between first floor and ground.

5.24 A rotational spring-damper system (Figure P5.24)
has K = 2 N·m and B = 6 N·m·s.   The applied angular
velocity function is:

$$\omega(t) = 8u_s(t) - u_r(t - 4) + u_r(t - 20), \text{ rad/s.}$$

Find:

(a) The angular velocity, $\omega_1(t)$.

(b) The torque in the damper after 8 s [i.e., $T_B(8)$].

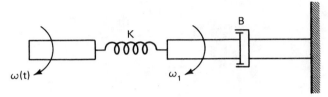

Figure P 5.24

CHAPTER 6

RESPONSE OF SECOND ORDER CIRCUITS

In the previous Chapter, we considered circuits with
only one storage element. These circuits could all be
described by a first-order differential equation and a
first-order characteristic equation. Since the characteristic equation had a single root, the natural response of
first-order circuits was limited to a single form, $Ae^{-t/\tau}$.

In this Chapter, we consider circuits with two storage
elements. Analysis of these circuits leads to second-
order differential equations and a quadratic characteristic
equation. Thus the characteristic equation always has two
roots for second-order circuits. The natural response may
now have any of three forms, one of which is generally
similar to the natural response of first-order circuits,

the other two being significantly different. The form
that is appropriate for a particular circuit depends on
the roots of the characteristic equation.

When the general input g(t) is operated on by a second-
order circuit the output function y(t) is produced. The
most general differential equation for second-order circuits is:

$$\frac{d^2y}{dt^2} + 2\delta\omega_N \frac{dy}{dt} + \omega_N^2 y = a_1 g(t) + a_2 \frac{dg(t)}{dt} + a_3 \frac{d^2g(t)}{dt^2} \quad (6.1)$$

where δ, ω_N, a_1, a_2 and a_3 are constants that depend on
the values of the circuit elements. The constant, δ, is
called the "damping ratio" and ω_N is called the "undamped
natural frequency".

The effect of δ and ω_N on the response will be made
clear when we derive the various forms of the natural
response.

Let us begin our consideration of two storage element
circuits, however, by suggesting a procedure for placing
a desired output variable in the general form given in
Equation (6.1).

6.1 FORMULATION OF DIFFERENTIAL EQUATIONS

We will limit our discussion here to circuits with two
storage elements and only one dissipative element. Of
course, the circuit laws we use are general and could be
applied to circuits of any complexity. However, we place
this restriction now to reduce the algebraic manipulation
required in the formulation of the circuit differential
equations. Circuits with more components are more conveniently treated by the impedance method which will be
presented in Chapter VII.

Consider, as a first illustration, the series RLC circuit
shown in Figure 6.1. There is one current (T-type) output
variable, i(t) and three potential (A-type) output variables, $e_R(t)$, $e_L(t)$, and $e_C(t)$. If we use the concept of

nodal potential, we can reduce the number of unknown potentials to two, $e_1(t)$ and $e_2(t)$. Then the three original
potential differences can be expressed in terms of the
nodal potentials and source potential. Thus:

$$e_R = e - e_1, \quad e_L = e_1 - e_2, \quad \text{and} \quad e_C = e_2$$

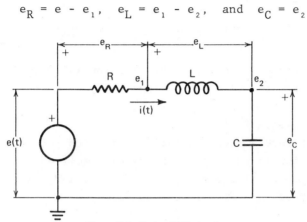

Figure 6.1 Series RLC circuit

We are left, therefore, with the variables $i(t)$, $e_1(t)$
and $e_2(t)$ which we wish to relate to the input, $e(t)$,
through a differential equation. The best way to accomplish this is to follow these general steps:

a) Select the type of variable that has the fewer
   unknowns. In this case, the variable is $i(t)$
   the single T-type variable. (There are two A-type
   variables, $e_1(t)$ and $e_2(t)$).

b) If the variable selected in part a) is T-type,
   begin by writing a compatibility equation (KVL).
   If the variable is A-type write a continuity equation
   (KCL). For the series RLC we therefore write the
   compatibility equation

$$[e(t) - e_1] + [e_1 - e_2] + [e_2] = e(t)$$

c) Substitute the elemental equations into the equation
   obtained in part b). Thus:

$$R\ i + L\ \frac{di}{dt} + \frac{1}{C} \int i\ dt = e(t)$$

d) Perform the mathematical operations required to put
the equation in part c) into the form given in
Equation (6.1)

$$\frac{d^2 i}{dt^2} + \frac{R}{L}\frac{di}{dt} + \frac{1}{LC}\ i = \frac{1}{L}\frac{de(t)}{dt}$$

Now that we have found the differential equation
that relates $i(t)$ to the input $e(t)$ we may easily
express the other outputs $e_1(t)$ and $e_2(t)$ in differential equation form.

e) Substitute a single elemental equation into the
equation in part c). For example, to find e_2 use
the elemental equation, $e_2 = (1/C) \int i\ dt$ or
$i = C\ de_2/dt$. The result is:

$$\frac{d^2 e_2}{dt^2} + \frac{R}{L}\frac{de_2}{dt} + \frac{1}{LC}\ e_2 = \frac{1}{LC}\ e(t)$$

To find e_1, substitute $i = [e(t) - e_1]/R$ into part
c) or part d) and obtain

$$\frac{d^2 e_1}{dt^2} + \frac{R}{L}\frac{de_1}{dt} + \frac{1}{LC}\ e_1 = \frac{1}{LC}\ e(t) + \frac{d^2 e(t)}{dt^2}$$

Note that the coefficients of the terms on the left-
hand side of the differential equations for the outputs
$i(t)$, $e_1(t)$ and $e_2(t)$ are the same. That is, the coefficient of the first derivative term is R/L and the coefficient of the third term is $1/LC$. Other circuits, in
general, will have other coefficients. However, for any
given particular circuit the differential equations for
all the circuit variables will have the same coefficients.
The coefficients are a characteristic of the circuit and
do not depend on the output variable.

To further demonstrate the procedure for formulating
the circuit differential equation consider the circuit

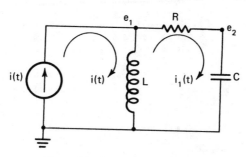

Figure 6.2 Parallel-Series RLC circuit

shown in Figure 6.2. The circuit has an inductance branch in parallel with a series RC branch. The input signal is from a current source, $i(t)$. Suppose we wish to find the differential equation that relates the potential $e_1(t)$ to the input source, $i(t)$. The steps to follow are:

a) Select the variable with fewer unknowns. There are two nodal variables e_1 and e_2. There is only one mesh variable $i_1(t)$. Thus, we select the variable, $i_1(t)$ and find its differential equation before the differential equation for $e_1(t)$.

b) Compatibility relation. Since the variable in part a) is $i_1(t)$ we write the compatibility equation around the mesh

$$- e_1 + (e_1 - e_2) + e_2 = 0$$

c) Substitute the elemental equations

$$- L \frac{d}{dt} (i(t) - i_1) + R\, i_1 + \frac{1}{C} \int i_1 \, dt = 0$$

d) Perform the mathematical operations:

$$\frac{d^2 i_1}{dt^2} + \frac{R}{L} \frac{di_1}{dt} + \frac{1}{LC} i_1 = \frac{d^2 i(t)}{dt^2}$$

e) Substitute the elemental equation for L in part d) and simplify. Thus:

$$i_1 = i(t) - \frac{1}{L} \int e_1 \; dt$$

and we obtain

$$\frac{d^2e_1}{dt^2} + \frac{R}{L}\frac{de_1}{dt} + \frac{1}{LC} e_1 = R \frac{d^2i}{dt^2} + \frac{1}{C}\frac{di}{dt}$$

We will follow this formulation procedure in the examples throughout this Chapter.

6.2 NATURAL RESPONSE OF SECOND ORDER CIRCUITS

The homogeneous form of the general differential equation for second-order circuits given in Equation (6.1) is:

$$\frac{d^2y_H}{dt^2} + 2\delta\omega_N \frac{dy_H}{dt} + \omega_N^2 \; y_H = 0 \qquad (6.2)$$

This is the general equation without the forcing functions and produces the unforced or natural response. As in the case of first-order circuits we assume a trial solution of the form $y_H = A \; e^{rt}$. If this solution is substituted into Equation (6.2) we obtain:

$$[r^2 + 2\delta\omega_N r + \omega_N^2]A \; e^{rt} = 0 \qquad (6.3)$$

Once again the trial solution is not equal to zero so that the bracketed term in Equation (6.3) must be equal to zero and

$$r^2 + 2\delta\omega_N r + \omega_N^2 = 0 \qquad (6.4)$$

Equation (6.4) is the general characteristic equation of second-order circuits. If we use the quadratic formula to solve Equation (6.4) for r the result is:

$$r = - \; \delta\omega_N \pm \omega_N \; \sqrt{\delta^2 - 1} \qquad (6.5)$$

Equation (6.5) represents the roots of the characteristic equation. The form of the natural response depends on whether the roots are real and distinct, real and equal, or a complex conjugate pair. This, in turn, depends on the value of δ, the damping ratio. From Equation (6.5) observe that when:

a) $\delta > 1$, the roots are real and distinct
b) $\delta < 1$, the roots are a complex conjugate pair
c) $\delta = 1$, the roots are real and equal

Each of these conditions produces a different form for the natural response and is treated separately.

a) $\delta > 1$

The natural response under this condition is sometimes called the "overdamped" response since large amounts of damping or resistance in circuits lead to large values of δ. In this case the characteristic equation has two real and distinct roots, r_1 and r_2. The natural response is similar to that of a first-order circuit and has the form:

$$y_H(t) = A_1 e^{r_1 t} + A_2 e^{r_2 t} \qquad (6.6)$$

Note that the sum of the roots, $r_1 + r_2$, will always equal $-2\delta\omega_N$ the negative of the coefficient of r in the characteristic equation.

The constants A_1 and A_2 cannot be evaluated until we obtain the complete solution and use the initial conditions. However, Equation (6.6) shows that the natural response will be the sum or difference of decaying exponentials.

b) $\delta < 1$

When the damping ratio is less than unity the response is termed "underdamped". This condition occurs when a small amount of damping or resistance is present in the

second-order circuit. The characteristic equation has
complex conjugate roots that are designated from Equation
(6.5) as:

$$r_1 = -\alpha + j\omega_d \qquad (6.7a)$$

$$r_2 = -\alpha - j\omega_d \qquad (6.7b)$$

where α, the damping constant, equals $\delta\omega_N$, and ω_d, the
damped natural frequency equals $\omega_N \sqrt{(1 - \delta^2)}$. The natural
response for the underdamped circuit is then:

$$y_H(t) = A_1 \exp[-(\alpha - j\omega_d)t] + A_2 \exp[-(\alpha + j\omega_d)t] \quad (6.8)$$

Equation (6.8) involving the complex exponentials is
inconvenient to work with. We may place the equation
in a more usable form by applying the Euler relation,
$e^{j\theta} = \cos\theta + j \sin\theta$. Thus Equation (6.8) may be reduced
to:

$$y_H(t) = e^{-\alpha t}[(A_1 + A_2)\cos\omega_d t + j(A_1 - A_2)\sin\omega_d t] \quad (6.9)$$

The natural response $y_H(t)$ described by Equation (6.9)
must be a real quantity for all values of time. Remember
that it represents the solution of a real physical problem.
For this condition to be satisfied the constants A_1 and A_2
must be complex conjugates. Then their sum will be a real
number and their difference is imaginary. It is therefore, convenient to express the natural response of an
underdamped second order circuit as:

$$y_H(t) = e^{-\alpha t} [A_3 \cos\omega_d t + A_4 \sin\omega_d t] \quad (6.10)$$

Figure 6.3 shows some typical underdamped natural responses for decaying sine and cosine functions. Note
that the responses are contained between an envelope of
decaying exponentials ($e^{-0.1t}$ and $-e^{-0.1t}$). For the

cases shown $\alpha = 0.1$ s^{-1} and $\omega_d = 1.0$ rad/s. Thus the
damping ratio here is $\delta = 0.0995$ and the natural frequency
$\omega_N = 1.005$ rad/s. When the damping ratio is larger (but
still less than 1.0) the enveloping exponentials decay
more rapidly and the magnitude of the peaks decreases.

c) $\underline{\delta = 1}$

When the roots are real and equal the system is "critically damped". We may treat this as the limiting case of
the overdamped response in which the two real roots
approach each other. Suppose we designate the roots
as r_1 and $r_1 + \Delta r$. Then the natural response may be
written as:

$$y_H(t) = \lim_{\Delta r \to 0} [A_1 \exp(r_1 t) + A_2 \exp(r_1 + \Delta r)t] \quad (6.11)$$

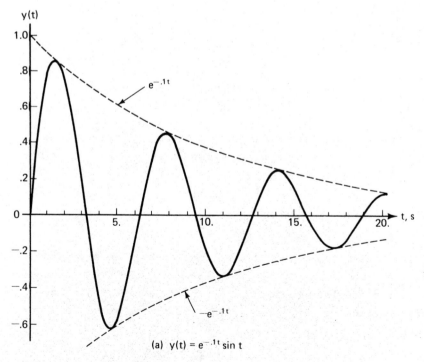

(a) $y(t) = e^{-.1t} \sin t$

Figure 6.3 Some typical underdamped natural responses ($\delta = 0.0995$, $\omega_N = 1.005$ rad/s)

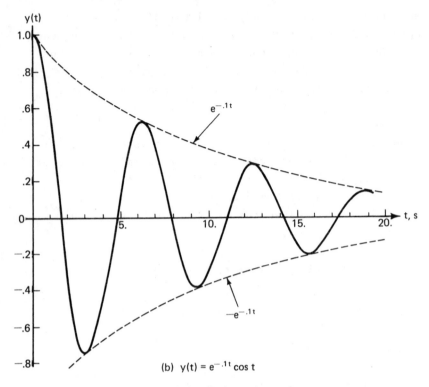

(b) $y(t) = e^{-.1t} \cos t$

Figure 6.3 Some typical underdamped natural responses
($\delta = 0.0995$, $\omega_N = 1.005$ rad/s) (cont'd)

If we add and subtract $A_2 \, e^{r_1 t}$ to the right-hand side
of Equation (6.11), we may express it as:

$$y_H(t) = (A_1 + A_2) \exp(r_1 t)$$

$$+ \lim_{\Delta r \to 0} [A_2 (\exp(r_1 + \Delta r)t - \exp(r_1 t))] \quad (6.12)$$

The constants, A_1 and A_2, are arbitrary. We may select
$A_5 = A_1 + A_2$ and $A_6 = A_2 \cdot \Delta r$. When A_5 and A_6 are evaluated
from the complete response it can be shown that neither
are functions of Δr. Thus with this choice we may express
Equation (6.12) as:

$$y_H(t) = A_5 \exp(r_1 t) + A_6 \lim_{\Delta r \to 0} \left[\frac{\exp(r_1 + \Delta r)t - \exp(r_1 t)}{\Delta r} \right] \quad (6.13)$$

Now from the definition of a derivative we recognize that:

$$\frac{d(e^{r_1 t})}{dr_1} = \lim_{\Delta r \to 0} \left[\frac{\exp(r_1 + \Delta r)t - \exp(r_1 t)}{\Delta r} \right] = t\, e^{r_1 t} \quad (6.14)$$

Thus, the homogeneous solution for the critically damped condition reduces to:

$$y_H(t) = (A_5 + A_6 t)e^{r_1 t} \quad (6.15)$$

Another way to approach the critically damped case is to assume a homogeneous solution of the form $y_H(t) = h(t)e^{r_1 t}$. Then the substitution of this form into Equation (6.2) shows that $d^2 h/dt^2$ must equal zero and therefore $h(t) = A_5 + A_6 t$.

Figure 6.4 compares the natural responses of a typical critically damped circuit and an equivalent overdamped

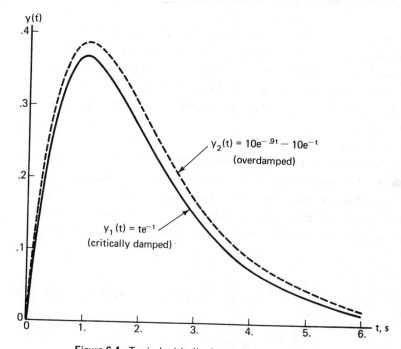

Figure 6.4 Typical critically damped natural response

circuit in which the roots are close together. The responses are nearly the same. There is never a significant
difference in appearance between an overdamped response
with damping ratio near unity and a critically damped
response.

Underdamped or critically damped responses may only occur
in circuits that have one T-type storage device and one
A-type storage device. The oscillatory or underdamped response is the result of an internal energy transfer between
an A-type and a T-type element. Circuits that contain only
one type of element (i.e., two T-types or two A-types)
must always be overdamped. For the damping ratio to be
equal to or less than unity elements of both types must
be present. However, the presence of both elemental types
is merely a necessary condition and does not ensure an
underdamped or critically damped response.

6.3 STEP RESPONSE OF SECOND ORDER CIRCUITS

As in the case of first order circuits we can find the
responses of second-order circuits by a mathematical
approach and a physical approach. The difference between
the two approaches is more evident for second-order circuits. To obtain the step response the input function,
$g(t) = G u_s(t)$.

a) Mathematical Approach.

In the mathematical approach we first formulate the
differential equation. The next step is to decompose
the equation so that only one term appears as a forcing
function on the right-hand side. For the general second
order circuit Equation (6.1), the decomposition results
in:

$$\frac{d^2 y_1}{dt^2} + 2\delta\omega_N \frac{dy_1}{dt} + \omega_N^2 y_1 = G u_s(t) \qquad (6.16a)$$

$$\frac{d^2 y_2}{dt^2} + 2\delta\omega_N \frac{dy_2}{dt} + \omega_N^2 y_2 = G u_i(t) \qquad (6.16b)$$

$$\frac{d^2 y_3}{dt^2} + 2\delta\omega_N \frac{dy_3}{dt} + \omega_N^2 \, y_3 = G \, u_d(t) \qquad (6.16c)$$

$$y_s(t) = a_1 \, y_1 + a_2 \, y_2 + a_3 \, y_3 \qquad (6.16d)$$

From the principles of linear differential equations, we may relate the solutions for y_2 and y_3 to the solution for y_1. In effect y_1 is the solution of a particular equation for a step input forcing function; y_2 and y_3 are solutions of the same equation for an impulse and doublet function respectively. This means that:

$$y_2 = \frac{dy_1}{dt} \qquad (6.17a)$$

$$y_3 = \frac{d^2 y_1}{dt^2} \qquad (6.17b)$$

and the complete solution for the general equation may be expressed solely in terms of y_1. From Equations (6.17) and (6.16d), we obtain:

$$y_s(t) = a_1 \, y_1 + a_2 \frac{dy_1}{dt} + a_3 \frac{d^2 y_1}{dt^2} \qquad (6.18)$$

Thus, the complete step response, $y_s(t)$ depends only on y_1. The problem is reduced, therefore, to finding a solution for y_1 in Equation (6.16a) and then applying Equation (6.18). The solution for y_1 will always be the sum of a homogeneous and a particular solution. The homogeneous solution depends on the damping ratio, δ, as indicated in the previous section. The particular solution of Equation (6.16a) is the value of y after all the transients have disappeared (i.e., $dy_1/dt = 0$, $d^2 y_1/dt^2 = 0$). The solution for y_1 is then:

$$y_1(t) = y_H + \frac{G}{\omega_N^2} \qquad (6.19)$$

where the particular solution is G/ω_N^2. The homogeneous portion of Equation (6.19) contains two arbitrary constants which must be evaluated from the initial conditions, $y_1(0^+)$

and $dy_1/dt(0^+)$. At this point, we must remember that the
decomposed Equation (6.16a) alone does not represent the
general second-order circuit. The initial conditions for
the decomposed equation are always $y_1(0^+) = 0$ and
$dy_1/dt(0^+) = 0$. The initial conditions for the actual
circuit will generally be different and can be determined
directly as indicated in Section 4.3.

The mathematical solution may be summarized in the
following steps:

1) Formulation. Find the complete circuit differential
   equation by the methods given in Section 6.1.

2) Decomposition. Replace the right-hand side of the
   complete equation by the step magnitude, G. Determine the response, $y_1(t)$ to the constant forcing
   function using $y_1(0^+) = 0$ and $dy_1/dt(0^+) = 0$.

3) Addition. Apply Equation (6.18) and obtain the complete step response.

b) Physical Approach

In the physical approach we also begin with the formulation of the differential equation. However, in this
approach, we really are only interested in finding the
characteristic equation. From the characteristic equation
we determine the homogeneous solution. The procedure
is identical to the physical approach used for first-order
circuits and is itemized as:

1) Homogeneous Solution. This solution depends on the
   value of the damping ratio, δ, and contains two
   arbitrary constants.

2) Particular Solution. The particular solution for a
   step input is the value of the output variable at
   infinite time. Thus $y_P = y_s(\infty)$

3) Initial Conditions. These conditions are found from
   a consideration of initial circuit performance during
   a sudden change (Section 4.3). They are not necessarily equal to zero. For the second-order circuit
   we need $y_s(0^+)$ and $dy_s/dt(0^+)$.

4) <u>Complete Solution</u>. Combine the homogeneous and particular solutions to form the complete solution. Use the initial conditions to determine the arbitrary constants.

We will demonstrate the solution procedure for the step response of second-order circuits with a few examples.

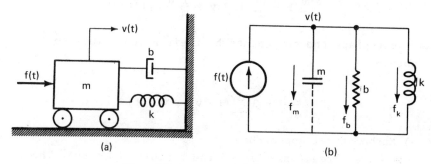

Figure 6.5 Mechanical circuit (example 6.1)

Example 6.1

The mechanical arrangement shown in Figure 6.5a is subjected to a driving force

$$f(t) = 10u_s(t), \text{ N}$$

If the mass, $m = 1$ kg, the damper, $b = 3$ N·s/m, and the spring, $k = 2$ N/m, find the velocity of the mass. Figure 6.5b shows the analogous circuit for the mechanical arrangement. The circuit consists of three elements in parallel and a T-type source (force source). To formulate the differential equation we use the steps given in Section 6.1. The variable type, v, (A-type) has the fewest unknowns and so we begin with a continuity equation:

$$f_m + f_b + f_k = f(t)$$

The substitution of the elemental equation yields

$$m \frac{dv}{dt} + bv + k \int v \, dt = f(t)$$

and by differentiation the differential equation for this circuit is:

$$\frac{d^2v}{dt^2} + \frac{b}{m} \frac{dv}{dt} + \frac{k}{m} v = \frac{1}{m} \frac{d}{dt} [f(t)]$$

We replace the components by their given values to obtain:

$$\frac{d^2v}{dt^2} + 3 \frac{dv}{dt} + 2v = 10u_i(t)$$

a) Mathematical Approach

We have already formulated the circuit equation. The next step is to place the equation in the decomposed form

$$\frac{d^2v_1}{dt^2} + 3 \frac{dv_1}{dt} + 2v_1 = 10$$

with

$$v_1(0^+) = 0$$

and

$$\frac{dv_1}{dt}(0^+) = 0 \quad \text{(by definition)}$$

If we compare the coefficients of this equation with the standard form (Equation (6.1)), we find that:

$$2\delta\omega_N = 3 \quad \text{and} \quad \omega_N^2 = 2$$

Thus, the damping ratio, $\delta = 1.06$, and the circuit is overdamped. The characteristic equation is:

$$r^2 + 3r + 2 = 0$$

and the roots of the equation are $r_1 = -1$ and $r_2 = -2$. The homogeneous solution is:

$$v_H = A_1 \, e^{-t} + A_2 \, e^{-2t}$$

The particular solution, $v_{1p} = 10/2 = 5$. The complete solution for $v_1(t)$ is:

$$v_1(t) = A_1 \, e^{-t} + A_2 \, e^{-2t} + 5$$

We now evaluate the arbitrary constants from the zero initial conditions and find that:

$$v_1(0^+) = A_1 + A_2 + 5 = 0$$

$$\frac{dv_1}{dt}(0^+) = -A_1 - 2A_2 = 0$$

with the result that $A_1 = -10$ and $A_2 = 5$. Thus, the complete solution for $v_1(t)$ is:

$$v_1(t) = [-10e^{-t} + 5e^{-2t} + 5]u_s(t)$$

This is not the velocity of the mass. To obtain the velocity of the mass we must use Equation (6.18) which in this case reduces to:

$$v(t) = \frac{d}{dt}[v_1(t)]$$

If we perform the differentiation

$$v(t) = (10e^{-t} - 10e^{-2t})u_s(t) + (-10e^{-t} + 5e^{-2t} + 5)u_i(t)$$

The second term on the right (the impulse term) is equal to zero for all values of time. Thus, the velocity of the mass is:

$$v(t) = [10e^{-t} - 10e^{-2t}]u_s(t)$$

b) Physical Approach

The homogeneous solution is in the same form as in the mathematical approach.

$$v_H = A_3 e^{-t} + A_4 e^{-2t}$$

We find the particular solution by inspection of the analogous circuit diagram (Figure 6.5b). After a long time the mass behaves like an open circuit and the spring like a short circuit. Thus:

$$v_p = v(\infty) = 0$$

The initial conditions are also determined from the circuit diagram noting that the mass is now a short and the spring is open. Thus immediately we know that $v(0^+) = 0$. To determine $dv/dt(0^+)$ we look for an elemental equation that contains dv/dt. For example, $f_m(t) = m\, dv/dt$. This equation holds for all values of time. If we specialize the equation for $t = 0^+$ the result is:

$$\frac{dv}{dt}(0^+) = \frac{f_m(0^+)}{m} = \frac{f(0^+)}{m} = 10$$

Since the mass behaves as a short all the force passes through this component at $t = 0^+$, and $f_m(0^+) = f(0^+)$.

The physical approach thus yields a complete solution

$$v(t) = A_3 e^{-t} + A_4 e^{-2t} + 0$$

and the arbitrary constants are determined from the initial conditions derived from the circuit performance

$$v(0^+) = A_3 + A_4 = 0$$

$$\frac{dv}{dt}(0^+) = -A_3 - 2A_4 = 10$$

from which $A_3 = +10$ and $A_4 = -10$. The complete step response for the velocity of the mass is:

$$v(t) = [10e^{-t} - 10e^{-2t}]u_s(t)$$

the same result that was obtained by the mathematical approach. Figure 6.6 shows the velocity of the mass. The velocity reaches a peak of 2.39 m/s when $t = 0.7$ s.

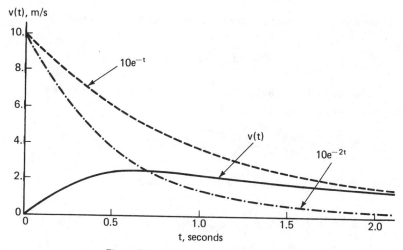

Figure 6.6 Velocity of mass (example 6.1)

Example 6.2

The electric circuit shown in Figure 6.7 has an input function

$$e_i(t) = u_s(t), \text{ V}$$

The component values are $R = 10$ Ω, $L = 0.592$ H, and $C = 0.01$ F. Find the value of the output voltage, $e_0(t)$.

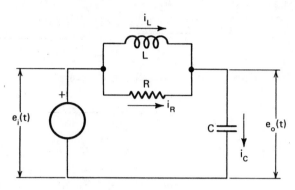

Figure 6.7   Electric circuit (example 6.2)

From the continuity equation at the output node:

$$i_c = i_R + i_L$$

If the elemental equations are used to replace the currents above the result is:

$$C \frac{de_o}{dt} = \frac{e_i - e_o}{R} + \frac{1}{L} \int (e_i - e_o) dt$$

Now by differentiation and rearrangement:

$$\frac{d^2 e_o}{dt^2} + \frac{1}{RC} \frac{de_o}{dt} + \frac{1}{LC} e_o = \frac{1}{LC} e_i(t) + \frac{1}{RC} \frac{d}{dt} [e_i(t)]$$

and with the component values

$$\frac{d^2 e_o}{dt^2} + 10 \frac{de_o}{dt} + 169 e_o = 169 u_s(t) + 10 u_i(t)$$

a) Mathematical Approach

In the decomposed form

$$\frac{d^2 e_{01}}{dt^2} + 10 \frac{de_{01}}{dt} + 169 e_{01} = 1$$

and

$$e_{01}(0^+) = 0, \quad \frac{de_{01}}{dt}(0^+) = 0, \quad \text{(by definition)}.$$

By comparison with the general second-order circuit
differential equation, the coefficients of the second
and third terms on the left-hand side yield:

$$2\delta\omega_N = 10 \quad \text{and} \quad \omega_N^2 = 169$$

From these relations we find that:

$$\omega_N = 13 \text{ rad/s and } \delta = 0.385$$

The circuit is underdamped and $\alpha = \delta\omega_N = 5 \text{ s}^{-1}$,
$\omega_d = \omega_N(1 - \delta^2)^{\frac{1}{2}} = 12 \text{ rad/s}$, and the homogeneous
solution is therefore:

$$e_H = e^{-5t} [A_1 \sin 12t + A_2 \cos 12t]$$

The particular solution, $e_p = 1/169$, (the value of e_{01}
when all the derivatives are set equal to zero indicating that the variable is no longer changing). The
form of the complete solution for e_{01} is:

$$e_{01}(t) = e^{-5t} [A_1 \sin 12t + A_2 \cos 12t] + 1/169$$

from the initial condition $e_{01}(0^+) = 0$, we find that
$A_2 = -1/169 = -0.00591$ and from $de_{01}/dt(0^+) = 0$,
$A_1 = (5/12) A_2 = -0.00246$.
   The complete solution for e_{01} is:

$$e_{01}(t) = e^{-5t} [-0.00246 \sin 12t - 0.00591 \cos 12t]$$

$$+ 0.00591$$

From Equation (6.18) the output voltage, $e_0(t)$ is
related to $e_{01}(t)$ by:

$$e_0(t) = 169 \, e_{01}(t) + 10 \, \frac{de_{01}(t)}{dt}$$

By substituting the expression for $e_{01}(t)$ in the
above equation we find that the output voltage is:

$$e_0(t) = [e^{-5t} (0.417 \sin 12t - \cos 12t) + 1]u_s(t)$$

This is the complete step response of the circuit.

b) Physical Approach

The homogeneous equation for this circuit is already known as:

$$e_H = e^{-5t} [A_3 \sin 12t + A_4 \cos 12t]$$

The particular solution for a step input is the final value of the variable. From the circuit diagram (Figure 6.7) the long-time performance after a step causes the inductor to act as a short and the capacitor to act as an open circuit. The final output voltage, therefore, equals the input voltage $e_0(\infty) = e_i(\infty) = 1$ V. The complete solution has the form:

$$e_0(t) = e^{-5t} [A_3 \sin 12t + A_4 \cos 12t] + 1$$

The arbitrary constants, A_3 and A_4, must be evaluated from the initial conditions in the circuit subject to a step input. For this circuit $e_0(0^+) = 0$ because the capacitor offers no resistance to a sudden change. The value of other initial conditions is obtained by evaluating the elemental equation for the capacitance at $t = 0^+$. Thus:

$$\frac{de_0}{dt}(0^+) = \frac{1}{C} i(0^+) = \frac{1}{RC} e_i(0^+) = 10$$

From these two initial conditions we find that $A_4 = -1$ and $A_3 = 0.417$ so that the complete solution is:

$$e_0(t) = [e^{-5t} [0.417 \sin 12t - \cos 12t] + 1]u_s(t)$$

which agrees with the previous result. Figure 6.8 shows the step response. The output voltage starts from zero and reaches 1.375 volts at t = 0.2 s. Thereafter, the voltage approaches the step value of 1 V in an oscillatory manner.

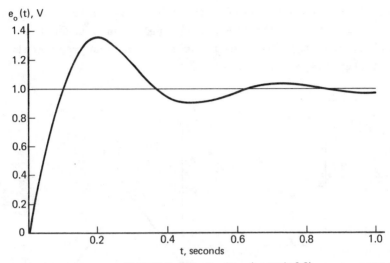

Figure 6.8 Output voltage (example 6.2)

6.4 RESPONSE OF SECOND-ORDER CIRCUITS TO OTHER INPUTS

We have developed the response of second-order circuits to excitation by a step function. Thus, as we have shown in Chapter V, for first-order circuits, we may use the properties of linear differential equations to determine the response for other inputs. The response of a linear circuit to inputs that are related to the step function by integration or differentiation can be obtained from the step response by integration or differentiation. To find the ramp or impulse response we may rewrite from Equations (5.28) and (5.32)

$$y_r(t) = \int_o^t y_s(t')dt' \qquad (6.20a)$$

$$y_i(t) = \frac{d}{dt} [y_s(t)] \qquad (6.20b)$$

When the input is a composite function we may separate
it into a sum of basic functions and find the composite
response by superposition of the basic function responses.
As a result, if we desire the response of a second-
order circuit to any specific input we first determine
its step response. Then we use the relation between the
specific input and the step input to operate appropriately
on the step response.

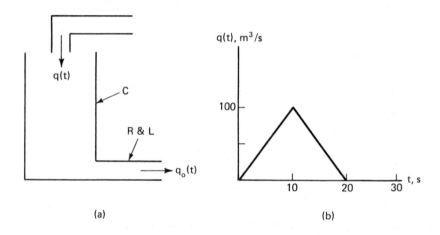

(a)   (b)

Figure 6.9 Fluid circuit (example 6.3)

Example 6.3

Water flows into a tank from above and then out through
a fluid line (Figure 6.9a). The capacitance of the tank,
$C = 0.1$ m^5/N, and the line has resistance, $R = 120$ N$\cdot$s/m^5
and inertance, $L = 1\ 000$ N$\cdot$s^2/m^5. Figure 6.9b shows the
flow rate $q(t)$ into the tank.

Find an expression for the flow rate out of the tank,
$q_0(t)$.

The analogous circuit for the tank and line combination is shown in Figure 6.10. The inlet flow rate is
a flow source. The tank represents a capacitance to

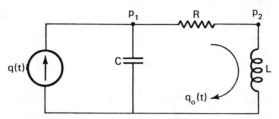

Figure 6.10 Analogous circuit (example 6.3)

ground and the line in this case is a series RL to ground. We designate the pressure at the bottom of the tank as p_1. The terminal, p_2, between R and L is fictitious in the sense that we cannot locate it physically. However, it is a convenient concept for the circuit analysis. A compatibility relation around the $q_0(t)$ mesh yields:

$$p_1 = (p_1 - p_2) + p_2 \quad \text{(KVLA)}$$

If the elemental equations are substituted into the compatibility relation the result is:

$$\frac{1}{C} \int [q(t) - q_0(t)] dt = R\, q_0 + L\, \frac{dq_0}{dt}$$

By differentiation and rearrangement we obtain the standard form:

$$\frac{d^2 q_0}{dt^2} + \frac{R}{L} \frac{dq_0}{dt} + \frac{1}{LC} q_0 = \frac{1}{LC} q(t)$$

The flow source, $q(t)$, may be described in terms of input functions as:

$$q(t) = 10u_r(t) - 20u_r(t - 10) + 10u_r(t - 20)$$

Thus when the elemental values are used the differential equation for this circuit and source is:

$$\frac{d^2q_0}{dt^2} + 0.12 \frac{dq_0}{dt} + 0.01 \, q_0$$

$$= 0.1u_r(t) - 0.2u_r(t - 10) + 0.1u_r(t - 20)$$

Comparison of the coefficients of this circuit equation with the standard form (Equation (6.2)) gives $2\delta\omega_N = 0.12$ and $\omega_N^2 = 0.01$. From this we find that $\omega_N = 0.1$ rad/s and $\delta = 0.6$. Thus, the circuit is underdamped.

We proceed now through the physical approach to find the step response. Since $\alpha = \delta\omega_N = 0.06$ s^{-1} and

$$\omega_d = \omega_N(1 - \delta^2)^{\frac{1}{2}} = 0.08 \text{ rad/s}$$

the homogeneous solution is:

$$q_H = e^{-0.06t} [A_1 \sin 0.08t + A_2 \cos 0.08t]$$

A long time after a unit step change the inertance is shorted and the capacitance is open. Under these circumstances all the source flow passes directly to the output and $q_p = q_0(\infty) = 1$.

The complete solution for the step input is:

$$q_{os}(t) = e^{-0.06t} [A_1 \sin 0.08t + A_2 \cos 0.08t] + 1$$

The initial conditions are now required to evaluate the arbitrary constants A_1 and A_2. Instantaneously, the capacitance is short and the inertance is open. Thus:

$$q_0(0^+) = 0$$

$$\frac{dq_0}{dt}(0^+) = \frac{P_2(0^+)}{L} = 0$$

With the use of these conditions, $A_1 = -0.75$ and $A_2 = -1$ so that the step response is:

$$q_{os}(t) = 1 - e^{-0.06t} [0.75 \sin 0.08t + \cos 0.08t]$$

To obtain the ramp response (Equation (6.20a)), we must integrate the step response between the limits $t' = 0$ to $t' = t$. Integration yields the unit ramp response as:

$$q_{or}(t) = t - 1.2 + 1.25 \ e^{-0.06t} \cos(0.08t + \phi)$$

where

$$\phi = 16.3° \quad or \quad 0.28 \ rad$$

This is the unit ramp response. However, recall that the input function is:

$$q(t) = 10u_r(t) - 20u_r(t - 10) + 10u_r(t - 20)$$

Thus by superposition and multiplication the response for the given input flow may be expressed as:

$$q_0(t) = 10[t-1.2+1.25e^{-0.06t} \cos(0.08t+0.28)]u_s(t)$$

$$- 20[t-11.2+1.25e^{-.06(t-10)}\cos(.08(t-10)+.28)]u_s(t-10)$$

$$+ 10[t-21.2+1.25e^{-.06(t-20)}\cos(.08(t-20)+.28)]u_s(t-20)$$

Figure 6.11 shows the output flow response and the input flow for this circuit. When the circuit is subjected to this input the output flow initially lags behind the input flow. Thus fluid is being stored in the tank. When the input is turned off ($t = 20$) there is still some fluid left in the tank. This fluid runs out at a slow rate, as shown.

Example 6.4

Figure 6.12 shows a simple mechanical circuit that consists of a mass, spring, and damper. The applied

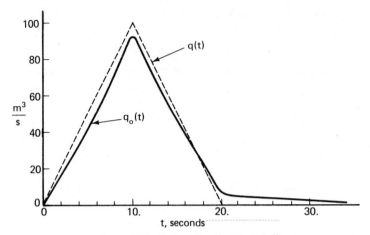

Figure 6.11 Output flow (example 6.3)

force, f(t), is a series of two impulses delivered by
a hydraulic ram. In terms of input functions, the
applied force, $f(t) = 1\ 000u_i(t) + 1\ 500u_i(t - 10)$, N.
If the mass m = 1 000 kg, the spring, k = 7.5 N/m and
the damper, b = 37.5 N·s/m, find the velocity of the
mass as a function of time.

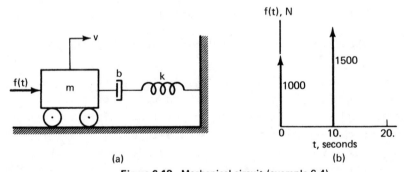

(a)   (b)

Figure 6.12 Mechanical circuit (example 6.4)

The analogous circuit for this example is the same as
the circuit in Example 6.3 (Figure 6.10). The usual
procedure for formulating the circuit equation yields:

$$\frac{d^2v}{dt^2} + \frac{k}{b}\frac{dv}{dt} + \frac{k}{m}v = \frac{k}{mb}f(t) + \frac{1}{m}\frac{df(t)}{dt}$$

For the particular elements given

$$\frac{d^2v}{dt^2} + 0.2\frac{dv}{dt} + 0.0075\,v = 0.2(10)^{-3}f(t) + (10)^{-3}\frac{df(t)}{dt}$$

A comparison of the coefficients with Equation (6.2) indicates that the damping ratio, $\delta = 1.155$ and the circuit is therefore overdamped. The characteristic equation of the circuit is:

$$r^2 + 0.2r + 0.0075 = 0$$

which has two real roots. The roots are $r_1 = -0.05$ and $r_2 = -0.15$. The homogeneous solution has the form:

$$v_H = A_1\,e^{-0.05t} + A_2\,e^{-0.15t}$$

If we use the physical approach, we must now find the final and the initial conditions for a step input. The final velocity depends only on the damper and the unit step force

$$v_P = v(\infty) = \frac{f(\infty)}{b} = 0.0267$$

The initial conditions are:

$$v(0^+) = 0, \quad \frac{dv}{dt}(0^+) = \frac{1}{m} = 0.001$$

The complete step response with arbitrary constants evaluated from the initial conditions is:

$$v_S(t) = 0.0033e^{-0.15t} - 0.03e^{-0.05t} + 0.0267$$

We use Equation (6.20b) and superposition to determine the response to the two impulses as:

$$v(t) = [1.5e^{-0.05t} - 0.5e^{-0.15t}]u_s(t)$$

$$+ [2.25e^{-0.05(t-10)} - 0.75e^{-0.15(t-10)}]u_s(t-10)$$

The response is shown plotted in Figure 6.13. The effects of the first impulse are still appreciable when the second impulse is applied. To find the distance moved by the mass, we would integrate the velocity response.

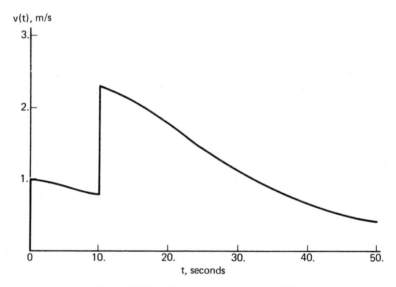

Figure 6.13 Velocity of mass (example 6.4)

6.5 RESPONSE OF SECOND-ORDER CIRCUITS WITH INITIAL STORAGE

Recall that when an A-type or T-type component has initial storage, we replace the component by a special circuit model. For A-type storage a potential source, $E_0 u_s(t)$, is placed in series with the A-type component. For T-type storage a flow source, $I_0 u_s(t)$, is placed in parallel with the T-type component.

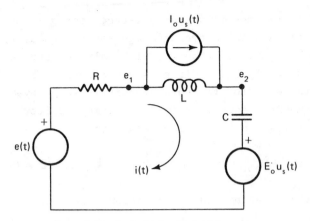

Figure 6.14 RLC circuit with initial storage

Figure 6.14 shows a series RLC circuit with initial storage in both capacitor and inductor. The resulting circuit has three sources. To obtain, for example, the current $i(t)$ we use the superposition of sources. We must, therefore, determine the effects of each source acting separately. Then:

$$i(t) = i_1(t) + i_2(t) + i_3(t)$$

where $i_1(t)$, $i_2(t)$, and $i_3(t)$ are the responses due to $e(t)$, $I_O u_s(t)$, and $E_O u_s(t)$ respectively. Figure 6.1 shows the appropriate circuit diagram to determine $i_1(t)$. Both storage sources have been suppressed. The differential equation for $i_1(t)$ has been derived in Section 6.1, and is:

$$\frac{d^2 i_1}{dt^2} + \frac{R}{L}\frac{di_1}{dt} + \frac{1}{LC} i_1 = \frac{1}{L}\frac{de(t)}{dt}$$

We must solve this equation by one of the methods previously given. The form of $e(t)$, of course, must be specified.

To find $i_2(t)$, the flow due to T-type storage, the input source, $e(t)$, and the A-type storage source, $E_O\,u_S(t)$ are suppressed. The circuit acting with only the current storage source, $I_O\,u_S(t)$, is shown in Figure 6.15a. The differential equation for this circuit is found by inserting the component equations into the compatibility relation $-e_1 + (e_1 - e_2) + e_2 = 0$. The result is:

$$Ri_2 + L\frac{d}{dt}\left[i_2 - I_O\,u_S(t)\right] + \frac{1}{C}\int i_2\,dt = 0$$

Now differentiation and rearrangement leads to:

$$\frac{d^2 i_2}{dt^2} + \frac{R}{L}\frac{di_2}{dt} + \frac{1}{LC}\,i_2 = I_O\,u_d(t)$$

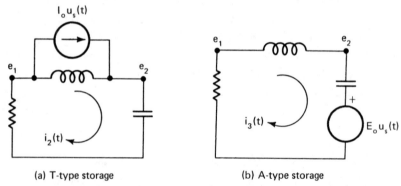

(a) T-type storage   (b) A-type storage

Figure 6.15 Storage sources acting alone in RLC circuit

The equation representing $i_2(t)$ has the same characteristic form as the equation for $i_1(t)$. In all cases, the characteristic equation of the circuit will not depend on the storage. However, when storage is present, the input or driving function will be specified by the product of the magnitude of the storage and some input function or functions.

The circuit acting with only the voltage storage source, $E_O\,u_S(t)$, is shown in Figure 6.15b. From the compatibility relation $-e_1 + (e_1 - e_2) + e_2 = 0$ and the component relations we have

$$Ri_3 + L \frac{di_3}{dt} + \frac{1}{C} \int i_3 \, dt + E_0 \, u_s(t) = 0$$

or

$$\frac{d^2 i_3}{dt^2} + \frac{R}{L} \frac{di_3}{dt} + \frac{1}{LC} i_3 = - \frac{1}{L} E \, u_i(t)$$

The input in this case is specified as $-E_0 \, u_i(t)/L$. Now we could solve the differential equations for i_1, i_2, and i_3 by the mathematical approach. However, it is much more convenient to use the physical approach. This requires the determination of the initial and final conditions. For storage type sources the final condition is always zero. We might expect this from a practical viewpoint since the stored energy is limited and eventually must run out. The following examples will demonstrate the treatment of second order circuits with storage.

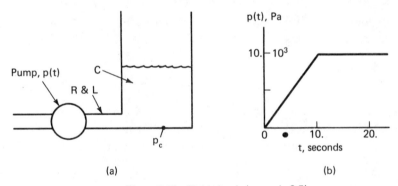

(a)   (b)

Figure 6.16 Fluid circuit (example 6.5)

Example 6.5

Figure 6.16a shows a pump, p(t) supplying water to a tank through a fluid line. The tank has a capacitance $C = 0.167$ m^5/N and initially is partially full of water so that the initial pressure at the bottom of the tank is 5 000 Pa. The line has a resistance $R = 50$ N·s/m^5 and an inertance $L = 100$ N·s^2/m^5. When the pump is turned on it takes 10 seconds to develop its full pres-

sure.   The pump pressure-time relation (p(t) vs. t) is
shown in Figure 6.16b).   Find the pressure at the bottom
of the tank as a function of time ($p_c(t)$).

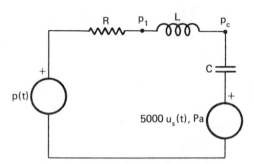

Figure 6.17   Analogous circuit (example 6.5)

Figure 6.17 shows the analogous circuit.   There are
two sources, (1) the pump and (2) the initial A-type
storage.   To find, $p_c(t)$ we must sum the responses of
the sources acting alone.   Thus:

$$p_c(t) = p_{c_1}(t) + p_{c_2}(t)$$

where p_{c_1} is due to the pump and p_{c_2} is due to the
initial storage.

The homogeneous equation for this circuit has been
derived as:

$$\frac{d^2 p_c}{dt^2} + \frac{R}{L}\frac{dp_c}{dt} + \frac{1}{LC} p_c = 0$$

and for the components given the characteristic equation
is:

$$r^2 + 0.5r + 0.06 = 0$$

The circuit is overdamped with roots of -0.2 and -0.3.

Pump Contribution (Storage source suppressed)
a) Unit Step Response
(i)   Homogeneous Solution, $p_H = A_1 e^{-0.2t} + A_2 e^{-0.3t}$

(ii) Final Conditions, $p(\infty) = 1$

(iii) Initial Conditions, $p(0^+) = 0$, $\frac{dp}{dt}(0^+) = 0$

(iv) Step Response, $p_{1s}(t) = -3e^{-0.2t} + 2e^{-0.3t} + 1$

b) Response to p(t)
   Now $p(t) = 1\ 000u_r(t) - 1\ 000u_r(t - 10)$. Thus, we
must integrate the step response and multiply it by 1 000.
The result is:

$p_{c_1}(t) =$

$$[1\ 000t - 8\ 333 + 15\ 000e^{-.2t} - 6\ 670e^{-.3t}]u_s(t)$$

$$- [1\ 000(t-10) - 8\ 333 + 15\ 000e^{-.2(t-10)}$$

$$-6\ 670e^{-.3(t-10)}]u_s(t-10)$$

Storage Source Contribution (Pump source suppressed)
a) Unit Step Response
(i)   Homogeneous Solution, $p_H = A_3 e^{-0.2t} + A_4 e^{-0.3t}$

(ii)   Final Condition, $p(\infty) = 0$
       This is always zero for storage sources.

(iii) Initial Conditions, $p(0^+) = 1$, $\frac{dp}{dt}(0^+) = 0$

(iv)   Step Response, $p_{2s}(t) = 3e^{-0.2t} - 2e^{-0.3t}$

b) Response to 5 000 $u_s(t)$

$$p_{c_2}(t) = [15\ 000e^{-0.2t} - 10\ 000e^{-0.3t}]u_s(t)$$

Tank Pressure

We may sum the contributions from the pump and the storage source to obtain:

$$p_c(t) =$$

$$[1\ 000t-8\ 333+30\ 000e^{-.2t}-16\ 670e^{-.3t}]u_s(t)$$

$$-\ [1\ 000(t-10)-8\ 333+15\ 000e^{-.2(t-10)}$$

$$-6\ 670e^{-.3(t-10)}]u_s(t-10)$$

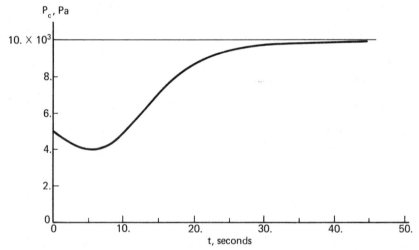

Figure 6.18 Tank pressure (example 6.5)

Figure 6.18 shows the tank pressure $p_c(t)$. The initial pressure is 5 000 Pa. Immediately after the pump is turned on the pressure decreases. This is a consequence of the pump characteristic which does not come up to rated pressure for 20 seconds. Thus, initially the pressure at the bottom of the tank is greater than the pump pressure and water will run out of the tank. At about 5 seconds the pump begins to take effect and water is pumped back into the tank. The pressure continues to rise thereafter until the pressure at the bottom of the tank equals the pump pressure.

Example 6.6

The flywheel in the rotational mechanical circuit shown in Figure 6.19a has a moment of inertia J = 50 kg·m² and is initially turning clockwise at 10 rad/s. The spring with constant K = 2 N·m has an initial counterclockwise torque of 50 N·m. The damper, B = 20 N·m·s/rad. Find the velocity of the flywheel as a function of time, ($\omega(t)$).

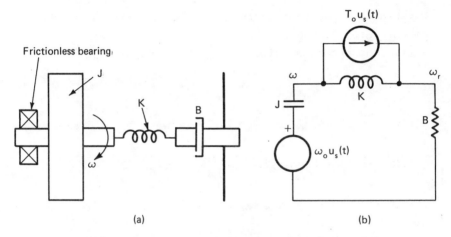

(a)   (b)

Figure 6.19 Rotational mechanical circuit (example 6.6)

Figure 6.19b shows the analogous circuit. There are two sources: (i) the initial A-type storage of the flywheel, $10u_s(t)$, and (ii) the initial T-type storage of the spring, $50u_s(t)$. The angular velocity response will be the sum of the responses due to each source.

$$\omega(t) = \omega_1(t) + \omega_2(t) .$$

where $\omega_1(t)$ and $\omega_2(t)$ are the responses of the A-type and T-type sources respectively when acting alone. The homogeneous equation for this circuit has the form:

$$\frac{d^2T}{dt^2} + \frac{K}{B}\frac{dT}{dt} + \frac{K}{J}T = 0$$

For the components in this circuit the characteristic equation is:

$$r^2 + 0.1\, r + 0.04 = 0$$

from which we find that the circuit is underdamped with $\delta = 0.25$ and $\omega_N = 0.2$ rad/s. Therefore, $\alpha = 0.05$ s^{-1} and $\omega_d = 0.194$ rad/s.

Flywheel Storage Contribution (Spring source suppressed)
a) Unit Step Response
    (i)   Homogeneous Solution, $\omega_H = e^{-\alpha t}[A_1\ \sin\omega_d t$

$$+\ A_2\ \cos\omega_d t]$$

    (ii) Final Condition, $\omega(\infty) = 0$ (always for storage)

    (iii) Initial Conditions, $\omega(0^+) = 1$, $\frac{d\omega}{dt}(0^+) = 0$

    (iv) Unit Step Response,

$$\omega_{1s}(t) = e^{-.05t}[.258\sin.194t + \cos.194t]$$

b) Response to $10u_s(t)$, (clockwise)

$$\omega_1(t) = e^{-0.05t}\,[2.58\ \sin\ 0.194t + 10\ \cos\ 0.194t]u_s(t)$$

Spring Storage Contribution (Flywheel source suppressed)
a) Unit Step Response
    (i)   Homogeneous Solution,

$$\omega_H = e^{-\alpha t}[A_3\ \sin\omega_d t + A_4\ \cos\omega_d t]$$

    (ii) Final Conditions, $\omega(\infty) = 0$ (always for storage)

    (iii) Initial Conditions, $\omega(0^+) = 0$, $\frac{d\omega}{dt}(0^+) = \frac{1}{J} = 0.02$

(iv) Unit Step Response,

$$\omega_{2s}(t) = -[0.103\sin.194t]e^{-.05t}$$

b) Response to $50u_s(t)$, (counterclockwise)

$$\omega_2(t) = [-5.15e^{-0.05t} \sin 0.194t]u_s(t)$$

Angular Velocity

We sum the contributions. The polarity in the circuit diagram already includes the proper directions. Thus:

$$\omega(t) = e^{-0.05t} [-2.57 \sin 0.194t + 10 \cos 0.194t]u_s(t)$$

$$= 10.32e^{-0.05t} \cos(0.194t + \phi)$$

where

$$\phi = 14.4°$$

Figure 6.20 shows the angular velocity response.

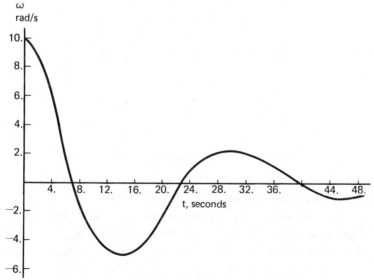

Figure 6.20   Angular velocity of flywheel (example 6.6)

Problems

6.1 The mass-spring system shown in Figure P6.1 is subjected to a horizontal impulsive force of $100u_i(t)$, N. The mass is 200 kg and the spring constant is 3 200 N/m. If the mass is initially at rest, find the velocity of the mass as a function of time.

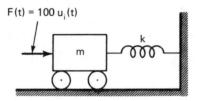

$F(t) = 100\, u_i(t)$

k

m

Figure P 6.1

6.2 A constant pressure pump is suddenly started to supply water to the fluid circuit shown in Figure P6.2. The circuit consists of two resistive lines and two tanks.

(a) Find the characteristic equation of the circuit.

(b) Prove that the damping factor, δ, must always be greater than unity.

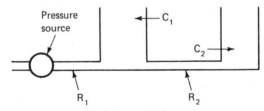

Pressure source

C_1

C_2

R_1

R_2

Figure P 6.2

6.3 The rotary mechanical system shown in Figure P6.3 is subjected to an input torque of $1\,350u_s(t)$, N·m. The moment of inertia of the flywheel $J = 30$ kg·m^2, the spring constant $K = 30$ N·m and the rotary damper $B = 15$ N·m·s. Express the angular velocity of the flywheel as a function of time.

Bearing (no friction)

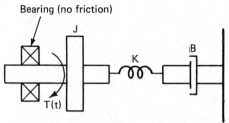

Figure P 6.3

6.4 Wind acting on the tail of a model rocket in flight acts as a torque input of $2(10)^{-3}u_s(t)$, N·m. This torque causes the rocket to rotate about an axis perpendicular to its length (pitch). The rocket can be modelled as a second-order system with $J = 4(10)^{-4}$ kg·m^2, $B = 2.53(10)^{-3}$ N·m·s, and K = 0.4 N·m. The damping is caused by air resistance and the spring effect is due to a restoring tendency. The damping and spring are in parallel. Find the angular pitch velocity, $\omega(t)$, and $\omega(0.25)$.

6.5 The mechanical system shown in Figure P6.5 is initially at rest. The mass, m = 10 kg, the spring, k = 20 N/m and the damper b = 20/3 N·s/m. Find the velocity of the mass, v(t) if the applied input force f(t) is:
  (a) $f(t) = 10u_s(t)$, N
  (b) $f(t) = 10u_r(t)$, N
  (c) $f(t) = 10u_i(t)$, N
  (d) $f(t) = 20u_s(t) - 10u_s(t - 2)$, N

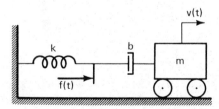

Figure P 6.5

6.6 In the mechanical system for Problem 6.5 the spring constant is changed to k = 10 N/m and the damper is changed to b = 25/3 N·s/m. Find v(t) if:

    (a) $f(t) = 10u_r(t) - 10u_r(t - 1)$, N

    (b) $f(t) = 10u_s(t - 2) + 10u_s(t - 3)$, N

6.7 For the rotational system shown in Figure P6.7,
$J = 4.5$ kg·m^2, $B = 22.5$ N·m·s, and $K = 27$ N·m/rad.
The angular velocity input $\omega(t) = 50u_s(t)$, rad/s.
Find the angular velocity of the flywheel and the
torque in the damper as functions of time.

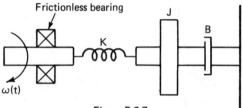

Figure P 6.7

6.8 A mass-spring-damper arrangement is shown in Figure
P6.8. The mass, m = 50 kg, the spring k = 200 N/m
and the damper b = 200 N·s/m. The spring is initially stretched 2.025 m and then the mass is released
from this position. The angle of the incline is 30
degrees. Find the velocity of the mass along the
incline as a function of time.

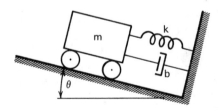

Figure P 6.8

6.9 For the mechanical system shown in Figure P6.9,
m = 25 kg, k = 200 N/m and b = 100/3 N·s/m. The
velocity input $v(t) = 5u_i(t) - 5u_i(t - 2)$, m/s.
Find:

    (a) the velocity of the mass as a function of time,
        $v_m(t)$

(b) the position of the mass as a function of time, $x_m(t)$

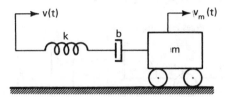

Figure P 6.9

6.10 A water tank has two outlet lines and one inlet line. One of the outlet lines is modelled as a pure resistance, $R = 12.5$ N·s/m^5 and the other as a pure inertance, $L = 16.67$ N·s^2/m^5. The inlet line has negligible resistance and inertance. It contains a flow pump which delivers q $u_r(t)$, m^3/s. The capacitance of the tank $C = 0.02$ m^5/N. Find q if the flow through the resistive line $q_R(2) = 31.29$ m^3/s.

6.11 The electrical circuit shown in Figure P6.11 has $L = 0.01$ H, $C = 4$ µF, $R = 100$ ohms, and $i(t) = 0.01$ $u_s(t)$, A. Find $i_R(t)$ and $e_L(t)$.

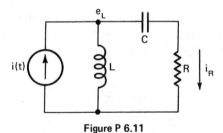

Figure P 6.11

6.12 A pneumatic system (Figure P6.12) consists of a resistive line, an inertive line and a bellows capacitance. All pressures are initially at atmospheric pressure when the pump is suddenly turned on and produces a pressure input of $(10)^{-4}$

$u_S(t)$ Pa.   If $R = 0.096(10)^{10}$ N·s/m^5, $C = 50(10)^{-10}$
m^5/N and $L = 0.32(10)^{10}$ N·s^2/m^5, find the rate of
flow through the inertive line.

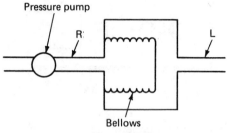

Figure P 6.12

6.13 A water tank is initially filled with water when
     two outlet lines are suddenly opened. The capaci-
     tance of the tank is $C = 0.25$ m^5/N. One outlet
     line is resistive and the other is inertive.
     Measurements show that the flow through the inert-
     ive line may be expressed as:

$$q_L(t) = 4\ 000\ [e^{-0.5t} - e^{-2t}]u_S(t),\ m^3/s$$

     Find the initial pressure at the bottom of the
     tank.

6.14 In the electric circuit shown in Figure P6.14
     $e(t) = 100u_S(t)$, V, $C = 0.2$ μF, $L = 0.3125$ H, and
     $R = 781.25$ ohms. Find $e_1(t)$ and $i_R(t)$.

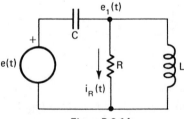

Figure P 6.14

6.15 In the rotary mechanical system shown in Figure
     P6.15 the spring ($K = 82$ N·m/rad) has been initially

turned through a clockwise angle of 3.5 radians.
The flywheel (J = 9 kg·m²) is subjected to a clockwise external torque of 315u$_p$(t), N·m. If the
damper B = 81 N·m·s, find the angle (θ_J(t)) through
which the flywheel turns.

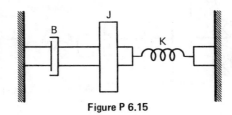

Figure P 6.15

6.16 An electric circuit consists of the parallel combination of L = 0.133 H and C = 5 μF connected in
series with a resistance (R = 80 ohms) to ground.
The input is a voltage source, e(t), shown in
Figure P6.16. Find the voltage across the resistance as a function of time.

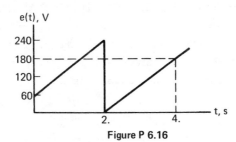

Figure P 6.16

6.17 In a parallel RLC electric circuit the capacitor
has an initial charge of 15(10)$^{-6}$ coulombs and the
inductor has an initial current of 0.1 A. If
C = 0.25 μF, L = 0.5 H and R = 667 ohms, find the
charge on the capacitor as a function of time.

6.18 For the rotational system shown in Figure P6.18
J = 16 kg·m², B = 135.2 N·m·s and K = 676 N·m/rad.
Initially with the clutch disengaged the spring is
twisted through an angle of 1 radian clockwise and
the flywheel is rotating with an angular velocity

of 6 rad/s counterclockwise. Find the angular
velocity of the flywheel as a function of time
after the clutch has been engaged.

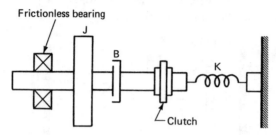

Figure P 6.18

6.19 A crude pressure transducer can be constructed by
attaching a strain gage to the flexible member
covering a cylindrical chamber. A test is then
performed to determine the response of the transducer. The test set-up consists of a long line
$(R = 7(10)^6$ N·s/m^5, $L = 2(10)^5$ N·s^2/m^5) leading to
the cylindrical chamber $(C = 167(10)^{-10}$ m^5/N). The
line is connected to a shock tube (Figure P6.19)
which generates a step in pressure of $200u_S(t)$, Pa.
Find an analytical expression for the pressure in
the chamber as a function of time.

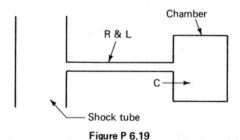

Figure P 6.19

6.20 A car is travelling along a straight road with
constant forward velocity when it hits a small
bump. The transverse velocity imparted to the car

by the bump is shown in Figure P6.20. The car
suspension is modelled as a mass sitting atop the
parallel combination of a damper and spring. If
m = 1 500 kg, b = 6 000 N·s/m and k = 60 000 N/m,
find $v_m(t)$, the transverse velocity of the mass.

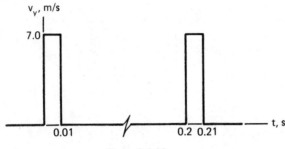

Figure P 6.20

6.21 Two compartments of a container are perfectly
insulated from the outside room. There is a
barrier between the compartments and it has a
thermal resistance of 0.02°C/W. Both compartments
are filled with water that is initially at 20°C.
The smaller compartment has a thermal capacitance
of $(10)^5$ J/°C and the larger compartment has a
thermal capacitance of $2(10)^5$ J/°C. A heating element in the smaller compartment is suddenly turned
on and supplies 3 000$u_s(t)$, W. Find the temperature
in each compartment when t = 1 000 s.

CHAPTER 7

GENERALIZED IMPEDANCES AND SYSTEM FUNCTIONS

   The model of any physical system with energy storage
results in a differential equation. Thus the determination
of system response requires a solution of a differential
equation. This holds for both the mathematical and physical
approaches given in Chapters V and VI. We may ask if it
is possible to model a system by an algebraic equation from
which the differential equation can be derived. To answer
this question we must consider the effect of the general
exponential input function. This type of driving function
in its most general mathematical form is not normally
encountered by a system. However, this function has two
special cases (step function and sine wave) which represent
the two most common driving functions used in practice.

7.1 THE GENERAL EXPONENTIAL INPUT FUNCTION

The most general form for the exponential function is given in Equation (2.25). If the time domain is restricted to t > 0 we may express the exponential function as:

$$g(t) = Ge^{st} u_s(t) \qquad (7.1)$$

where, in general, s, is a complex number, $\sigma + j\omega$. In complex form Equation (7.1) may be written as:

$$g(t) = Ge^{\sigma t}[\cos\omega t + j \sin\omega t]u_s(t) \qquad (7.2)$$

The real Re[ ] and imaginary Im[ ] parts of the general exponential function are therefore:

$$Re[g(t)] = Ge^{\sigma t} \cos\omega t\ u_s(t) \qquad (7.3a)$$

$$Im[g(t)] = Ge^{\sigma t} \sin\omega t\ u_s(t) \qquad (7.3b)$$

The shape of the input signals given in Equation (7.3) depends on the values of σ and ω. Figure 7.1 shows the function for four possible combinations of σ and ω. The most common special cases are the step input (Figure 7.1a) and the sine wave (Figure 7.1c). Less well known are the decaying exponential (Figure 7.1b) and the decaying sine wave (Figure 7.1d). As a result, if we can find the response of a given system to the general exponential function (Equation 7.1,, we may, by specification of s, obtain the response for all the special cases. Although we do not specify the form of s in the following responses, the examples given in this chapter will consider the effect of a purely real value. We will reserve the speci-fication of s as a pure imaginary value until Chapter VIII, when we discuss frequency response.

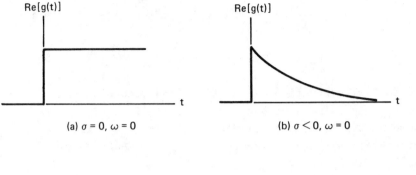

(a) $\sigma = 0$, $\omega = 0$

(b) $\sigma < 0$, $\omega = 0$

(c) $\sigma = 0$, $\omega =$ constant

(d) $\sigma < 0$, $\omega =$ constant

Figure 7.1 Some exponential functions

7.2 RESPONSE TO EXPONENTIAL FUNCTION

First-Order Circuits

The differential equation for the general first-order circuit has been given in Equation (5.1) and is:

$$\tau \frac{dy}{dt} + y = a_1 \, g(t) + a_2 \, \frac{dg(t)}{dt} \qquad (7.4)$$

where in this case $g(t) = Ge^{st} u_s(t)$. We may obtain a complete solution for Equation (7.4) by decomposing the equation in the same manner as in Equation (5.22). The result is:

$$\tau \frac{dy_1}{dt} + y_1 = a_1 \, Ge^{st} u_s(t) \qquad (7.5a)$$

$$\tau \frac{dy_2}{dt} + y_2 = a_2 G \frac{d}{dt} [e^{st} u_s(t)] \qquad (7.5b)$$

$$y = y_1 + y_2 \qquad (7.5c)$$

We deal with Equation (7.5a) first and recognize that $y_1 = y_{1H} + y_{1P}$. The homogeneous solution is $A_1 e^{-t/\tau}$. We assume a form for the particular solution, $A_2 e^{st}$ and then evaluate A_2 by substituting this form into Equation (7.5a). This procedure yields

$$\tau s A_2 e^{st} + A_2 e^{st} = Ga_1 e^{st} \qquad (7.6)$$

from which we find that $A_2 = a_1 G/(\tau s + 1)$ and therefore:

$$y_{1P} = \frac{a_1 G}{\tau s + 1} e^{st} \qquad (7.7)$$

The solution for $y_1(t)$ can now be formulated through the application of the initial condition $y(0^+) = 0$ and is:

$$y_1(t) = \frac{a_1 G}{\tau s + 1} [e^{st} - e^{-t/\tau}] u_s(t) \qquad (7.8)$$

Now $y_2(t) = \frac{a_2}{a_1} \frac{dy_1}{dt}$ since the forcing function in Equation (7.5b) is a linear function of the derivative of the forcing function in Equation (7.5a). Thus:

$$y_2(t) = \frac{a_2 G}{\tau s + 1} [se^{st} + \frac{1}{\tau} e^{-t/\tau}] u_s(t) \qquad (7.9)$$

and the complete solution for Equation (7.4) subject to an exponential input is:

$$y(t) = G \left[\frac{a_1 + a_2 s}{\tau s + 1} e^{st} + \frac{\frac{a_2}{\tau} - a_1}{\tau s + 1} e^{-t/\tau} \right] u_s(t) \qquad (7.10)$$

When $s = 0$, the condition where the general exponential function becomes a step function, Equation (7.10) reduces to the form given for the step response (Equation (5.24)).

Example 7.1

A force $f(t) = F e^{st} u_s(t)$ is applied to the mass-damper system shown in Figure 7.2a. Obtain an expression for the resulting velocity.

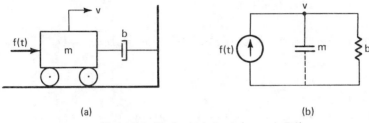

(a) (b)

Figure 7.2 Mechanical circuit (example 7.1)

The analogous circuit diagram for this mechanical arrangement is shown in Figure 7.2b. The equilibrium equation is found as before in Example 5.6 as:

$$\frac{m}{b} \frac{dv}{dt} + v = \frac{1}{b} f(t)$$

$$= \frac{F}{b} e^{st} u_s(t), \text{ for the forced response}$$

$$= 0, \text{ for the natural response}$$

The solution can be obtained directly from Equation (7.10) by making the substitutions $a_1 = 1/b$, $a_2 = 0$, $G = F$, and $\tau = m/b$. However, let us go through the solution process for this case. As always the natural response (homogeneous solution) is an exponential of the form $v_H(t) = A_1 e^{-t/\tau}$. In this system the time constant $\tau = m/b$. The forced response (particular solution) of any linear system has the same shape as the driving function or its derivatives. Thus, for the mass-damper system under consideration the forced response has the form:

$$v_p(t) = V e^{st}$$

where V is a constant that must be determined by substitution of the forced response form back into the
differential equation. If we perform this operation
the result is:

$$(ms + b)V = F \quad \text{or} \quad V = \frac{F}{ms + b}$$

Thus the complete solution for the velocity is:

$$v(t) = \frac{F}{ms + b} e^{st} + A_1 e^{-bt/m}$$

where A_1 must be evaluated by application of the initial
conditions. Since forces are constrained to finite
values the initial velocity of mass is the same as
observed with step driving functions. That is:

$$v(0^+) = v(0^-) = 0$$

if the mass is at rest before the force is applied. From
this condition we find that:

$$A_1 = - F/(ms + b)$$

and thus the complete solution for velocity is:

$$v(t) = \frac{F}{ms + b} [e^{st} - e^{-bt/m}]u_s(t)$$

This result is consistent with the step response of
the system as may be observed by substituting the value
$s = 0$.

For the case where the forcing function exponent,
$s = -1/\tau = -b/m$, the above solution appears to be
indeterminate $(0/0)$. However, through the use of
L'Hopital's rule we may show that:

$$v(t) = \lim_{s \to -b/m} \left[\frac{\frac{d}{ds}(F e^{st} - F e^{-bt/m})}{\frac{d}{ds}(ms + b)} \right]$$

and the solution is therefore:

$$v(t) = \frac{Ft}{m} e^{-bt/m}$$

This form may be recognized as a special case of a critically damped second-order circuit (Equation (6.15)). The response of a first-order circuit driven by an exponential function with $s = -1/\tau$ may be, therefore, indistinguishable from the response of a critically damped second-order circuit if there is no forcing function or the forced response to a step input is zero.

As a corollary of the discussion in Example 7.1, suppose we are given a response of the type:

$$y(t) = A_1 e^{\alpha t} + A_2 e^{\beta t} \tag{7.11}$$

where α and β are real negative. We are asked to determine what kind of a circuit produced such a response. There are actually three possibilities and we are unable to distinguish among them. The possibilities are:

(a) First-order circuit with time constant equal to $-1/\alpha$ and input exponential $e^{\beta t}$.
(b) First-order circuit with time constant equal to $-1/\beta$ and input exponential $e^{\alpha t}$.
(c) Second-order circuit that is overdamped and has no forcing input function or the forced response to a step input is zero.

Second-Order Circuits

In general, second-order circuits produce the differential equation given in Equation (6.1). This is:

$$\frac{d^2 y}{dt^2} + 2\delta\omega_N \frac{dy}{dt} + \omega_N^2 y = a_1 g(t) + a_2 \frac{dg(t)}{dt} + a_3 \frac{d^2 g(t)}{dt^2} \tag{7.12}$$

If we apply an exponential input for $g(t)$ and decompose the resulting equation so that $y = y_1 + y_2 + y_3$ we obtain:

$$\frac{d^2 y_1}{dt^2} + 2\delta\omega_N \frac{dy_1}{dt} + \omega_N^2 y_1 = a_1 G\, e^{st}\, u_s(t) \quad (7.13a)$$

$$\frac{d^2 y_2}{dt^2} + 2\delta\omega_N \frac{dy_2}{dt} + \omega_N^2 y_2 = a_2 G\, \frac{d}{dt}\, [e^{st}\, u_s(t)] \quad (7.13b)$$

$$\frac{d^2 y_3}{dt^2} + 2\delta\omega_N \frac{dy_3}{dt} + \omega_N^2 y_3 = a_3 G\, \frac{d^2}{dt^2}[e^{st}\, u_s(t)] \quad (7.13c)$$

The forcing function in Equation (7.13b), is the derivative of the forcing function in Equation (7.13a). Similarly, the forcing function in Equation (7.13c) is the second
derivative of the one in Equation (7.13a). As a result,
the complete solution can be expressed in terms of y_1 as:

$$y = a_1 y_1 + a_2 \frac{dy_1}{dt} + a_3 \frac{d^2 y_1}{dt^2} \quad (7.14)$$

where y_1 is the response to $G\, e^{st}\, u_s(t)$.

We now proceed to find the solution for y_1 as the sum
of the homogeneous solution y_{1H} and the particular solution
y_{1P}. As we have observed in Chapter VI, the homogeneous
solution may take one of three forms depending on whether
the system is overdamped, underdamped, or critically
damped. To determine the particular solution we must
substitute $y_{1P} = A\, e^{st}\, u_s(t)$ into Equation (7.13a) and
then evaluate A. This procedure leads to:

$$[s^2 + 2\delta\omega_N s + \omega_N^2]A = G \quad (7.15)$$

The solution for y_1 is therefore:

$$y_1(t) = \frac{G\, e^{st}}{s^2 + 2\delta\omega_N s + \omega_N^2} + y_H \quad (7.16)$$

where $y_{1H} = y_H$ since there is only one homogeneous solution for each specific circuit. That is, $y_{1H} = y_{2H} = y_{3H}$
$= y_H$. The homogeneous solution contains two arbitrary
constants. These constants may be evaluated from Equation

(7.16) by applying the initial conditions $y_1(0^+) = 0$ and $dy_1/dt\ (0^+) = 0$. After the constants have been determined the complete solution is obtained by substituting Equation (7.16) into Equation (7.14).

$$y(t) = \frac{G(a_1 + a_2 s + a_3 s^2)e^{st}}{s^2 + 2\delta\omega_N s + \omega_N^2} + a_1 y_H + a_2 \frac{dy_H}{dt} + a_3 \frac{d^2 y_H}{dt^2} \quad (7.17)$$

Equation (7.17) is merely meant to indicate the formulation of a solution. In specific cases each system should be solved without recourse to this general formulation.

7.3 IMPEDANCE OF BASIC COMPONENTS

It is important to notice that in Equations (7.10) and (7.17) the use of the exponential forcing function resulted in the polynomials $(\tau s + 1)$ and $(s^2 + 2\delta\omega_N s + \omega_N^2)$ appearing in the forced response. These polynomials had previously been encountered as $(\tau r + 1)$ and $(r^2 + 2\delta\omega_N r + \omega_N^2)$ in the characteristic equations of first-order systems (Equation (5.7)) and second-order systems (Equation (6.4)). The characteristic equations were used to obtain the natural responses of the systems. In each case the factor r appeared as a result of a differentiation and the factor $1/r$ appeared as a result of an integration. In the case of exponential forcing functions it should be evident that the factors s and $1/s$ appear for exactly the same reason. Thus it should not be surprising that the operations of differentiation and integration are expressible algebraically through the use of an operator. The choice of a symbol for the operator is quite arbitrary. We select the symbol s since it is the most common in system and circuit analysis.

It should be possible, therefore, to find a method of writing equilibrium equations which is equally suitable for obtaining the natural response, the forced response

or the complete response. We may accomplish this by
representation of the basic components in algebraic terms
rather than as time derivatives or integrals. The algebraic
replacement used to describe the basic circuit components
is called the "generalized impedance".

The generalized impedance of an element is defined as
the ratio of the amplitude of the exponential across
variable to the amplitude of the exponential through
variable. The generalized impedance, $Z(s)$, is a function
of the complex exponent, s and is expressed mathematically
as:

$$Z(s) = \frac{\text{Amplitude of an exponential across variable}}{\text{Amplitude of the corresponding through variable}} \quad (7.18)$$

To derive the impedances of the three basic components
we begin with the elemental relations given in Equation
(3.1). The relations are:

$$(A.V.) = K_1 \, (T.V.) \qquad \text{dissipative} \quad (7.19a)$$

$$(A.V.) = K_2 \, \frac{d}{dt} \, (T.V.) \qquad \text{T-type storage} \quad (7.19b)$$

$$(A.V.) = K_3 \, \int (T.V.) dt \qquad \text{A-type storage} \quad (7.19c)$$

Now we assume that both the across variable difference
and through variable can be represented by exponential
functions. That is:

$$A.V. = A \, e^{st} \qquad (7.20a)$$

$$T.V. = T \, e^{st} \qquad (7.20b)$$

When Equations (7.20) are substituted into Equations
(7.19) the exponential terms cancel and we are left with
a relationship between the amplitudes A and T. If we
then use the definition of generalized impedance as the
ratio of A/T we find that:

$$Z(s) = K_1 \quad , \quad \text{dissipative} \qquad (7.21a)$$

$$Z(s) = K_2 s \quad , \quad \text{T-type storage} \qquad (7.21b)$$

$$Z(s) = K_3/s \quad , \quad \text{A-type storage} \qquad (7.21c)$$

Equation (7.21) is therefore the algebraic representation of the basic components that we had previously described in Equation (7.19) with time differentials and integrals.

Table 7.1 lists the generalized impedance in the various physical media.

TABLE 7.1 GENERALIZED IMPEDANCES OF ELEMENTS

	Dissipative	A-Type	T-Type
Mechanical Translation	1/b	1/ms	s/k
Mechanical Rotation	1/B	1/Js	s/k
Electric Circuit	R	1/Cs	Ls
Fluid System	R	1/Cs	Ls
Thermal System	R	1/Cs	--

The reciprocal of the generalized impedance is often a more convenient way to represent the basic elements. This is called the "generalized admittance", $Y(s)$, and is defined as:

$$Y(s) = \frac{\text{Amplitude of an exponential through variable}}{\text{Amplitude of the corresponding across variable}} \qquad (7.22)$$

Before proceeding further it is important to acknowledge that traditionally the mechanical engineer has used the term "mechanical impedance" from a viewpoint of force; that is, the larger the force required to produce movement, the larger the mechanical impedance. With this premise,

"mechanical impedance" is given by F/V. Having chosen the
concept of the through/across variables as the basic
premise, this text uses the electric circuit concept of
resistance to the flow of current and generalizes it by
taking "generalized impedance" as the ratio V/F. In some
of the mechanical engineering literature the ratio V/F is
called the "mobility". However, it will be found much
simpler if the term "generalized impedance" is used exclusively to represent the ratio of across to through variable
and the term "generalized admittance" is used exclusively
for the ratio of through to across variable. This practice
is followed throughout this text.

Impedances in Combination

Generalized impedances and admittances can be combined
in much the same way that equivalent resistances are
obtained for series and parallel connections of resistors.
The essential difference is that the energy storage elements, being terms in s or 1/s, must have their impedances
or admittances added algebraically so that in general, an
equivalent impedance or admittance is a polynomial in s.
The determination of a series or parallel connection must
be made only by inspection of the analogous circuit. That
is, it is only when the analogous circuit shows unequivocally that the flow of the through variable is identically
the same for two or more elements that they may be
considered as a series circuit. Similarly, it is only
when the analogous circuit shows unequivocally that the
across variable is identically the same for two or more
elements that they may be considered as a parallel circuit.
   Figure 7.3a shows three impedances (Z_1, Z_2 and Z_3) interconnected so that there is a common through variable. This
is, therefore, a series circuit. Suppose we wish to represent the circuit by the circuit with equivalent single
impedance (Z_e) shown in Figure 7.3b. This requires that
the applied potential (AV_{TOTAL}) produce the same through

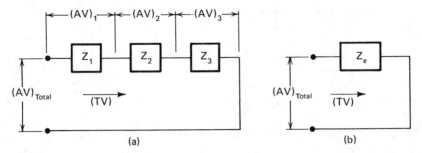

Figure 7.3 Impedances in series

variable in both circuits. The applied potential in Figure 7.3a may be expressed as:

$$(AV)_{TOTAL} = (AV)_1 + (AV)_2 + (AV)_3 \qquad (7.23)$$

or in terms of impedances (Equation (7.18)), Equation (7.23) is:

$$(AV)_{TOTAL} = Z_1(TV) + Z_2(TV) + Z_3(TV) \qquad (7.24)$$
$$= (Z_1 + Z_2 + Z_3)(TV)$$

The relation between the applied potential and through variable in the equivalent circuit (Figure 7.3b) is:

$$(AV)_{TOTAL} = Z_e(TV) \qquad (7.25)$$

A comparison of Equations (7.24) and (7.25) shows that the equivalent impedance for a series circuit is:

$$Z_e = Z_1 + Z_2 + Z_3 \qquad (7.26)$$

Let us now consider the equivalent impedance for a parallel circuit. Figure 7.4a shows three impedances (Z_1, Z_2, Z_3) that have a common across variable and are therefore in parallel. To replace this circuit by a circuit with an equivalent single impedance (Z_e), Figure

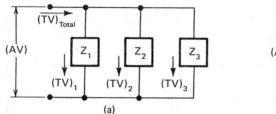 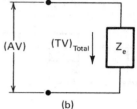

(a)   (b)

Figure 7.4 Impedances in parallel

7.4b, a through variable (TV_{TOTAL}) must produce the same across variable in each circuit. For the parallel circuit in Figure 7.4a:

$$(TV)_{TOTAL} = (TV)_1 + (TV)_2 + (TV)_3 \qquad (7.27)$$

and in terms of impedance Equation (7.27) becomes:

$$(TV)_{TOTAL} = \frac{AV}{Z_1} + \frac{AV}{Z_2} + \frac{AV}{Z_3} \qquad (7.28)$$

$$= (\frac{1}{Z_1} + \frac{1}{Z_2} + \frac{1}{Z_3})AV$$

For the equivalent circuit shown in Figure 7.4b:

$$(TV)_{TOTAL} = (\frac{1}{Z_e})AV \qquad (7.29)$$

From Equations (7.28) and (7.29) we may observe that the equivalent impedance for a parallel circuit is:

$$\frac{1}{Z_e} = \frac{1}{Z_1} + \frac{1}{Z_2} + \frac{1}{Z_3} \qquad (7.30)$$

Figure 7.5a shows admittances (Y_1, Y_2, Y_3) in series and Figure 7.5b shows an equivalent admittance (Y_e) representation. We may determine the relation between the equivalent admittance and the individual admittances by proceeding directly from the series circuit summation of across variables given in Equation (7.23). However, since we

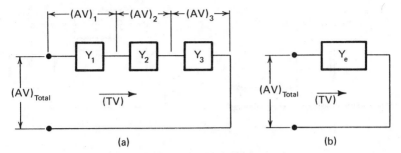

Figure 7.5 Admittances in series

know that admittance is the reciprocal of impedance we
may use the series circuit result given in Equation (7.26)
to obtain:

$$\frac{1}{Y_e} = \frac{1}{Y_1} + \frac{1}{Y_2} + \frac{1}{Y_3} \qquad (7.31)$$

In a similar way we may find the equivalent admittances
in parallel. Thus if we invert the equivalent impedance
and the individual impedances for the parallel circuit as
given in Equation (7.30) the result is:

$$Y_e = Y_1 + Y_2 + Y_3 \qquad (7.32)$$

Admittances in parallel are therefore additive just as
impedances in series.

Circuit components may be represented by either impedance
or admittance, whichever is more convenient for a particular circuit. Care must be exercised, however, to designate
clearly whether impedance or admittance is being used.

Example 7.2

A mass-damper system (Figure 7.6a) is driven by a
velocity input $v(t) = V\,e^{st}$. The resulting force through
the damper is $f(t) = F\,e^{st}$. Find the equivalent impedance, $Z_e = V/F$ for the system.

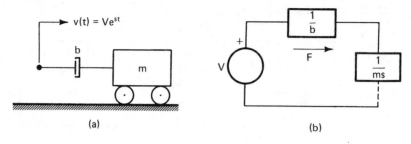

Figure 7.6 Mechanical circuit (example 7.2)

Figure 7.6b shows the impedance circuit diagram for
the system. The impedance of the damper is 1/b and of
the mass is 1/ms. The impedances are connected in a
simple series combination. Thus according to Equation
(7.26) the equivalent impedance is the sum of the individual impedances

$$Z_e = \frac{1}{b} + \frac{1}{ms} = \frac{b + ms}{bms}$$

Example 7.3

The position of the mass and damper in Example 7.2
are reversed (Figure 7.7a). In this arrangement the
system is driven by a force input $f(t) = F\,e^{st}$. The
resulting velocity of the mass is expressible as
$v(t) = V\,e^{st}$. Find the equivalent impedance $Z_e = V/F$
and the equivalent admittance $Y_e = F/V$ for the system.

The impedance circuit for this arrangement is shown in
Figure 7.7b. The components are in parallel. From

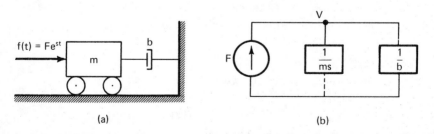

Figure 7.7 Mechanical circuit (example 7.3)

Equation (7.30) the equivalent impedance of a parallel circuit is:

$$\frac{1}{Z_e} = \frac{1}{Z_b} + \frac{1}{Z_m} = b + ms$$

or

$$Z_e = \frac{1}{b + ms}$$

The equivalent admittance can be found by inverting the equivalent impedance given above or from Equation (7.32). Thus:

$$Y_e = Y_b + Y_m = b + ms$$

Driving Point Impedance and Admittance

The ratio of across-variable amplitude to through variable amplitude as observed across the input terminals of a circuit is called the driving point impedance (Z_{dp}). We may obtain the driving point impedance by combining impedances in series and parallel until the entire circuit is represented by a single equivalent impedance. The driving point admittance is merely the reciprocal of the driving point impedance. The following examples will clarify the procedure.

Example 7.4

Figure 7.8a shows an impedance circuit with impedances Z_1, Z_2, Z_3, and Z_4. Find the driving point impedance between the terminals 1-2.

The procedure is to apply the rules of series and parallel circuits (Equations (7.26) and (7.30)) in sequence until only one equivalent impedance remains. Accordingly, we begin by combining the series connection of Z_3 and Z_4 into an equivalent impedance, $Z_3 + Z_4$. The circuit then appears as shown in Figure 7.8b. Now the

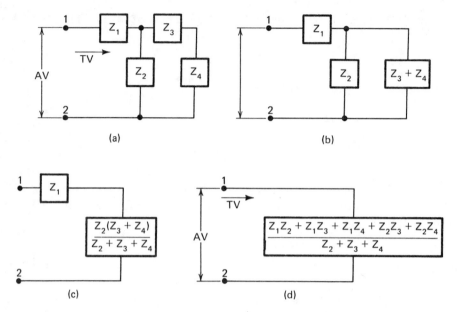

Figure 7.8 Determination of driving point impedance (example 7.4)

parallel combination of Z_2 and $Z_3 + Z_4$ produces an equivalent impedance (according to Equation (7.30)) of:

$$\frac{1}{Z_e} = \frac{1}{Z_2} + \frac{1}{Z_3 + Z_4}$$

or

$$Z_e = \frac{Z_2(Z_3 + Z_4)}{Z_2 + Z_3 + Z_4}$$

With this combination the circuit reduces to the simple series circuit shown in Figure 7.8c. Finally, the addition of the two impedances in series permits us to represent the circuit by a single impedance (Figure 7.8d). This is the driving point impedance

$$Z_{dp} = Z_1 + \frac{Z_2(Z_3 + Z_4)}{Z_2 + Z_3 + Z_4}$$

$$= \frac{Z_1 Z_2 + Z_1 Z_3 + Z_1 Z_4 + Z_2 Z_3 + Z_2 Z_4}{Z_2 + Z_3 + Z_4}$$

Example 7.5

Figure 7.9a shows a circuit in which the components have been represented by their admittances Y_1, Y_2, Y_3, and Y_4. Find the driving point admittance between the terminals 1-2.

We must apply the rules for combining admittances (Equations (7.31) and (7.32)). The series combinations of Y_1, Y_2 and Y_3, Y_4 are shown in Figure 7.9b. This follows directly from Equation (7.31). The two resulting admittances are in parallel and may be treated by direct addition (Equation (7.32)). The result is:

$$
\begin{aligned}
Y_e &= \frac{Y_1 Y_2}{Y_1 + Y_2} + \frac{Y_3 Y_4}{Y_3 + Y_4} \\
&= \frac{Y_1 Y_2 Y_3 + Y_1 Y_2 Y_4 + Y_1 Y_3 Y_4 + Y_2 Y_3 Y_4}{Y_1 Y_3 + Y_2 Y_3 + Y_1 Y_4 + Y_2 Y_4}
\end{aligned}
$$

7.4 SYSTEM FUNCTIONS

In the previous examples the ratio of the response to the excitation has been either a generalized admittance

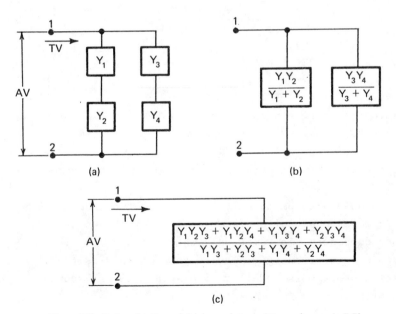

(a)

(b)

(c)

Figure 7.9 Determination of driving point admittance (example 7.5)

or a generalized impedance since a suitable through
variable was chosen as the response to an across variable
excitation, or a suitable across variable was chosen as
the response to a through variable excitation. In previous
chapters, there have been examples where this has not been
the case (e.g. excitation and response both across variables or both through variables). It is usual to define
a System Function, $G(s)$ for the more general situation

$$G(s) = \frac{\text{amplitude of exponential response}}{\text{amplitude of exponential excitation}} = \frac{\text{Output}}{\text{Input}} \quad (7.33)$$

The system function may therefore be a generalized admittance, a generalized impedance, the ratio of two across
variables, or the ratio of two through variables. The
general form is the ratio of two polynomials in s:

$$G(s) = \frac{b_m s^m + b_{m-1} s^{m-1} + \ldots \ldots + b_o}{a_n s^n + a_{n-1} s^{n-1} + \ldots \ldots + a_o} \quad (7.34a)$$

$$= \frac{(s-s_a)(s-s_b)(s-s_c) \ldots \ldots}{(s-s_1)(s-s_2)(s-s_3) \ldots \ldots (s-s_n)} \quad (7.34b)$$

where a_n, b_m, are the coefficients of the polynomials and
s_a, $s_b \ldots, s_1, s_2 \ldots$ refer to the roots of the polynomials.
   Although, strictly speaking, a system function is obtained for the forced response only, it is a simple matter
to "recover" the original differential equation by recalling the association between s and d/dt, and between $1/s$
and $\int dt$.
   It should also be noted that a system function implies
a single exponential source. As a result the system
function concept is valid only when the storage elements
have no initial energy. The reason for this is that
initial storage is modelled with an additional source
(Chapter IV).

There are two situations, the potential divider and the flow divider, which often appear in circuits and are readily treated with system functions. Figure 7.10 shows the potential and flow divider circuits. The potential divider circuit (Figure 7.10a) is basically a series circuit and has only a single through variable associated with it. The system function that relates the across variable output, $(AV)_o$, and the across variable input, $(AV)_i$, is:

$$G(s) = \frac{(AV)_o}{(AV)_i} = \frac{Z_2}{Z_1 + Z_2} \qquad (7.35a)$$

$$= \frac{Y_1}{Y_1 + Y_2} \qquad (7.35b)$$

where Equation (7.35b) represents the system function for the potential divider in terms of admittances.

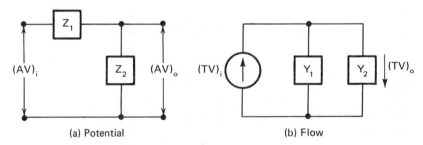

(a) Potential   (b) Flow

Figure 7.10 Potential and flow divider circuits

The flow divider circuit (Figure 7.10b) is a parallel circuit. There is only one across variable difference across both impedances. The system function for the ratio of the through variable output, $(TV)_o$, to through variable input, $(TV)_i$, is:

$$G(s) = \frac{(TV)_o}{(TV)_i} = \frac{Y_2}{Y_1 + Y_2} \qquad (7.36a)$$

$$= \frac{Z_1}{Z_1 + Z_2} \qquad (7.36b)$$

where Equation (7.36b) is the system function for the flow
divider circuit in terms of impedances.

Example 7.6

Figure 7.11a shows an RLC series circuit. The output
$e_o(t)$, is the voltage across the capacitance. The input
is supplied by an ideal voltage source, $e_i(t)$.

(a) Find the system function that relates the output
signal to the input signal.

(b) From the system function determine the differential
equation of the system.

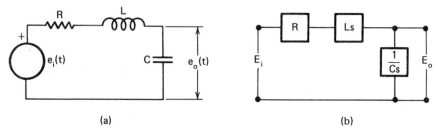

(a)   (b)

Figure 7.11   Electric circuit (example 7.6)

(a) It is convenient to redraw the circuit diagram
with impedance blocks rather than component symbols.
This is illustrated in Figure 7.11b. We may now place
this circuit in potential divider form by noting that
the resistance and inductance in series produce an
impedance, $R + Ls$. The system function now follows
directly from the application of the potential divider
Equation (7.35a). Thus:

$$\frac{E_o}{E_i} = \frac{\frac{1}{Cs}}{(R + Ls) + \frac{1}{Cs}} = \frac{1}{LCs^2 + RCs + 1}$$

(b) Cross multiplication of the system function yields:

$$LCs^2 E_o + RCs\, E_o + E_o = E_i$$

Now we may return to the time domain by recalling that s represents differentiation with respect to time. Therefore, we may write directly that:

$$LC \frac{d^2 e_o}{dt^2} + RC \frac{de_o}{dt} + e_o = e_i$$

Example 7.7

In a fluid system the system function that relates the output pressure, P_o, to the input pressure, P_i, is:

$$G(s) = \frac{P_o}{P_i} = \frac{s + 2}{s + 4}$$

If $p_i(t) = u_s(t)$, find $p_o(t)$ for this system.

The first step is to transfer the system function into the time domain and recover the governing differential equation. We accomplish this by cross-multiplying the system function so that:

$$s\, P_o + 4\, P_o = s\, P_i + 2\, P_i$$

This algebraic equation may be transferred now into the time domain with the result that:

$$\frac{dp_o}{dt} + 4\, P_o = \frac{dp_i}{dt} + 2\, P_i$$

Now we may solve the differential equation by decomposing as in Chapter V (Equation (5.22)) and Chapter VI. Thus:

$$\frac{dp_{o1}}{dt} + 4\, P_{o1} = 2\, P_i = 2u_s(t)$$

$$\frac{dp_{o2}}{dt} + 4\, P_{o2} = \frac{dp_i}{dt} = u_i(t)$$

$$P_o = P_{o1} + P_{o2}$$

When decomposition is used recall that the initial conditions on the separated equations are always zero (i.e. $p_{o_1}(0^+) = 0$).

The solution for p_{o_1} is therefore:

$$p_{o_1} = \frac{1}{2} (1 - e^{-4t})u_s(t)$$

We may determine the solution for p_{o_2} by noting that:

$$p_{o_2} = \frac{1}{2} \frac{d}{dt} (p_{o_1})$$

thus:

$$p_{o_2} = \frac{1}{2} [\frac{1}{2} 4e^{-4t}]u_s(t) = e^{-4t} u_s(t)$$

The required $p_o(t)$, is the sum of p_{o_1} and p_{o_2} so that:

$$p_o(t) = \frac{1}{2} (1 + e^{-4t})u_s(t)$$

When a system function is expressed as an output over an input, the characteristic equation of the system can be obtained by equating the denominator of the system function to zero. Thus from Equation (7.34a) the general characteristic equation of a system is:

$$a_n s^n + a_{n-1} s^{n-1} + \ldots \ldots a_o = 0 \qquad (7.37)$$

The roots of this equation, s_1, s_2 $\ldots$ s_n, are called the "poles" of the system function. Although excitation by an exponential function, e^{st}, where s is exactly equal to one of the poles (s_1, s_2 $\ldots$ s_n) would appear to produce an infinite response this is not necessarily the case. In Example 7.1, we observed that when the forcing exponential was equal to a root of the characteristic equation, the first-order system responded as a critically damped second-order system. In general, excitation of any system with a value of s equal to one of the system poles produces a dynamic response which has the same form as that of the next higher order system with two (or more) coincident poles. It may also be observed from Equation (7.34b) that excitation at any root of the numerator of the system function (s_a, s_b, s_c) produces zero forced response. These roots (s_a, s_b, s_c ...) are therefore called the "zeros" of the system function.

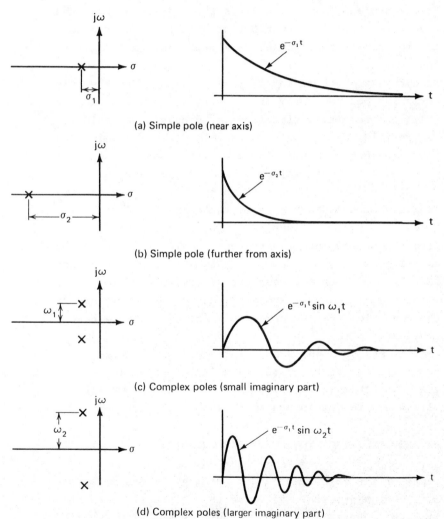

(a) Simple pole (near axis)

(b) Simple pole (further from axis)

(c) Complex poles (small imaginary part)

(d) Complex poles (larger imaginary part)

Figure 7.12 Correspondence of time response to pole positions

It is common to think of "poles" and "zeros" in graphical
terms by plotting their position on a complex plane. The
result is that the general form of the natural response of
a system may be associated directly with the position of
the poles on the complex s-plane. Figure 7.12 shows the
correspondence between pole location and time response.
In each case the figure on the left gives a pole position
on the s-plane (s = σ + jω) and the figure on the right
gives the resulting contribution of that pole arrangement
to the transient response. A simple pole (s = $-\sigma_1$) near
the jω axis as shown in Figure 7.12a leads to a slowly
decaying exponential in the time domain. When the pole
moves further away from the jω axis (s = $-\sigma_2$) as in Figure
7.12b the resulting time function decays more rapidly.
Figure 7.12c shows the complex poles (s = $-\sigma_1 \pm j\omega_1$) and
the corresponding damped sine wave response. When the
poles move further away from the real axis as in Figure
7.12d (s = $-\sigma_1 \pm j\omega_2$) the frequency of the response increases but the decay rate is the same.

For the physical circuits which emanate from the combination of elemental components (resistance, inductance,
capacitance), the system poles always lie on or to the
left of the jω axis. Thus the corresponding time response
either remains at a fixed amplitude or decays. This is
the condition that occurs in all the circuits we have
considered here. However, in some systems, (e.g. with
amplifiers), the poles may be located in the right half
plane (to the right of the jω axis). Such pole locations
give rise to time functions with increasing amplitudes and
ultimately result in unstable systems.

7.5 GENERAL FORMULATION OF SYSTEM EQUATIONS

One of the main reasons for seeking a circuit representation of a physical system is that it facilitates the
process of determining the circuit equilibrium equations.
Simple algorithms exist from which these equations can be

written, often directly from an inspection of the circuit
diagram. In this way the basic techniques of circuit
analysis can be applied to physical systems other than
electrical and the analogous properties of such systems
become quite evident.

Circuit analysis is based on the two Kirchhoff circuit
laws. These have been considered in Chapter IV (Equations
(4.1) and (4.2)) but are restated here in the form which
is most convenient for the analysis of more complicated
circuits than had previously been considered. For
Kirchhoff's Current Law (KCL):

$$\Sigma(TV)_P = \Sigma(TV)_S \qquad (7.38a)$$

Equation (7.38) means that the sum of all through variables leaving a junction through the passive elements
connected directly to the junction, $(TV)_P$, is equal to
the sum of all through variables entering the junction
from through variable sources, $(TV)_S$.

For Kirchhoff's Voltage Law (KVL):

$$\Sigma(AV)_P = \Sigma(AV)_S \qquad (7.38b)$$

which states that the sum of the across variable drops
across passive elements, $(AV)_P$, around a closed circuit
path, equals the sum of across variable increases due to
across variable sources, $(AV)_S$.

Up to this point, the most complicated system which has
been described in detail, by means of one or two equilibrium equations, has had only three elements. We are
now in a position to develop systematic procedures to
obtain sets of equilibrium equations for any system, no
matter how complicated, using its analogous circuit diagram
as the common intermediate step. The circuit diagram is
at first simply a graphical representation of the interactions which take place among all the elements of a system
in qualitative terms. However, each interaction can be

modelled by means of a simple elemental mathematical
relationship and thus the circuit effectively becomes a
graphical representation of the equilibrium equations.
The problem is therefore to find a method whereby the
information contained in the circuit diagram may be
quantified quickly and reliably in the form of equilibrium
equations which, in general, will be a set of simultaneous
differential equations. If we use the ideas developed in
this chapter, we will be able to derive the equations more
confidently and with less effort. We therefore imply that
all the driving functions are exponential, and use the
generalized impedance notation. It should be clearly
understood that this does not in any way limit the validity
or generality of the resulting equations.

Determination of Analogous Impedance Circuits

   Since the circuit diagram is the basis on which our
unified approach rests, it is appropriate to review the
process by which the circuit was obtained in previous
chapters. Let us consider the mass-spring-damper system
in Figure 7.13a.

   The first step is to identify all parts which are constrained to move at the same velocity and show the direction
chosen to be positive. Number these points since they will
form the junctions or nodes of the circuit diagram. In
this case, there are three independent velocities, v_1, v_2,
v_3 and a reference velocity, v_g. The equations are usually
easier to write and the results easier to interpret if
positive velocity is taken in the same direction at each
part of the system. Now draw the "skeleton" of the circuit,
i.e., the 3 nodes corresponding to the 3 parts of the
system at which the velocities are different and the datum
or reference node. In Figure 7.13b these are indicated
as V_1, V_2, V_3 and V_g. The impedance blocks corresponding
to the passive elements are added to noting the velocity
difference which is required in the force-velocity relation
of the element. The damper, b_1, models an effect dependent

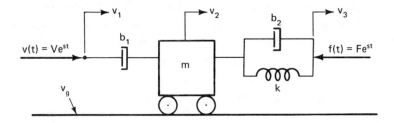

(a) Mechanical arrangement

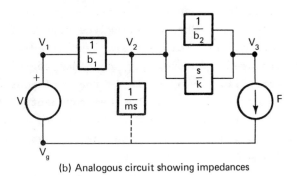

(b) Analogous circuit showing impedances

Figure 7.13 Determination of analogous circuit

on the relative velocity between nodes 1 and 2. Therefore, the impedance block $(1/b_1)$ is inserted between these nodes. The damper, b_2 and the spring, k, both depend on the relative velocity between nodes 2 and 3. Thus the damper block $(1/b_2)$ is connected in parallel with the spring impedance block (s/k) between nodes 2 and 3. The mass, on the other hand, models an effect dependent on the absolute velocity of node 2. To make this a two-terminal component we use the ground as a fictitious terminal and connect the mass impedance block $(1/ms)$ between V_2 and V_g.

Now, the branches having the sources can be added. The velocity source at node 1 constrains node 1 to move to the right at a velocity V relative to the datum. Therefore, on the circuit diagram this is represented by a velocity source connected between node 1 and the datum.

It is positive at node 1 because the source constrains
node 1 to move in the positive direction. The force source
acting at node 3 would have moved node 3 in the negative
direction if it had been acting separately. Thus the "flow
of force" into node 3 must also be negative and the arrow
is directed away from node 3.

Node-to-Datum Equation Formulation

The node-to-datum equations are usually referred to as
node equations. The method is a very convenient way of
formulating equations when the desired output is of the
across variable type. When applying the node-to-datum
method the coefficients turn out simplest if admittance
parameters are used. Accordingly, consider the circuit
shown in Figure 7.14. Suppose it is the analogous circuit
for a particular fluid system that has exponential flow
sources with amplitudes Q_1 and Q_2. We seek to find the
equilibrium equations that describe the exponential
pressure amplitudes P_1 and P_2.

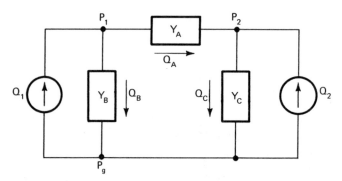

Figure 7.14 Analogous circuit involving admittances

We begin by applying Equation (7.38) (KCL) to nodes 1
and 2.

$$\text{Node } 1 \rightarrow Q_A + Q_B = Q_1 \tag{7.39a}$$

$$\text{Node } 2 \rightarrow Q_C - Q_A = Q_2 \tag{7.39b}$$

where flows away from the node are positive on the left-hand side of the equation and negative on the right-hand side. Once all the node equations are written (in this case the two Equations (7.39a) and (b)) we formulate the component relations. The three component relations are:

$$(P_1 - P_2)Y_A = Q_A \qquad (7.40a)$$

$$(P_1 - P_g)Y_B = Q_B \qquad (7.40b)$$

$$(P_2 - P_g)Y_C = Q_C \qquad (7.40c)$$

Now, if Equations (7.40) are substituted into Equations (7.39) and we note that the reference pressure, P_g, is zero, the result is:

$$(Y_A + Y_B)P_1 - (Y_A)P_2 = Q_1 \leftarrow \text{Node 1} \qquad (7.41a)$$

$$- (Y_A)P_1 + (Y_A + Y_C)P_2 = Q_2 \leftarrow \text{Node 2} \quad (7.41b)$$

Let us consider the individual terms in these two equations and try to interpret them.

Node 1: $(Y_A + Y_B)P_1$ = total flow away from node 1 due to P_1 only, all other pressures set to zero.

$- (Y_A)P_2$ = total flow away from node 1 due to P_2 only, all other pressures set to zero.

Q_1 = total flow into node 1 from all flow sources connected to node 1.

Node 2: $(Y_a + Y_C)P_2$ = total flow away from node 2 due to P_2 with all other variables set to zero.

$- Y_a P_1$ = total flow away from node 2 due to P_1 with all other variables set to zero.

Q_2 = total flow into node 2 from all flow
sources connected to node 2.

At this stage, it is valuable to note the coefficients
in the equations, and the matrix form of the equations is
useful for this purpose.

$$\begin{array}{c}\text{Node 1} \rightarrow \\ \\ \text{Node 2} \rightarrow \end{array}\begin{bmatrix} (Y_a + Y_b) & -Y_a \\ \\ -Y_a & (Y_a + Y_c) \end{bmatrix}\begin{bmatrix} P_1 \\ \\ P_2 \end{bmatrix} = \begin{bmatrix} Q_1 \\ \\ Q_2 \end{bmatrix} \quad (7.42)$$

In the array of admittances the terms on the main diagonal
are positive and all others are negative. This is true in
general for node-to-datum equilibrium equations.

The general form is:

$$\begin{bmatrix} Y_{11} & Y_{12} \\ \\ Y_{21} & Y_{22} \end{bmatrix}\begin{bmatrix} P_1 \\ \\ P_2 \end{bmatrix} = \begin{bmatrix} Q_1 \\ \\ Q_2 \end{bmatrix} \quad (7.43)$$

Note that Y_{11} is the sum of the admittances of the
passive branches connected to node 1 and Y_{22} is the sum
of the admittances at node 2. Y_{12} is the negative of the
admittance of the branch connecting nodes 1 and 2. If
every element connecting two nodes is "bilateral" (i.e.,
the pressure-flow relationship is the same for flow in
both directions) $Y_{12} = Y_{21}$ or $Y_{jk} = Y_{kj}$ and the matrix
is symmetrical.

With some practice, it is a simple matter to write the
equilibrium equations of many systems directly in this
form (i.e., "by inspection" of the system circuit) thus
reducing the likelihood of error. We will clarify the
procedure in the following two examples.

Example 7.8

The mechanical arrangement shown in Figure 7.15a consists of two masses, three dampers and one spring. There

are two known force inputs (f_1 and f_2). Present the node-to-datum equations in matrix form for this system.

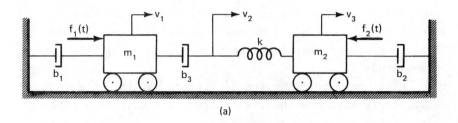

(a)

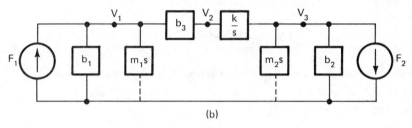

(b)

Figure 7.15 Mechanical circuit (example 7.8)

Figure 7.15b shows the circuit representation of the mechanical system. Since there are three unknown velocities the required matrix must be of the form:

$$
\begin{array}{cc}
& \begin{array}{cc} \downarrow & \downarrow \\ \text{unknowns} & \text{knowns} \end{array} \\
\begin{bmatrix} Y_{11} & Y_{12} & Y_{13} \\ Y_{21} & Y_{22} & Y_{23} \\ Y_{31} & Y_{32} & Y_{33} \end{bmatrix}
\begin{bmatrix} V_1 \\ V_2 \\ V_3 \end{bmatrix}
=
\begin{bmatrix} F_1 \\ 0 \\ -F_2 \end{bmatrix}
\begin{array}{l} \leftarrow \text{ node 1} \\ \leftarrow \text{ node 2} \\ \leftarrow \text{ node 3} \end{array}
\end{array}
$$

$$
\begin{array}{ccc}
\uparrow & \uparrow & \uparrow \\
V_2 = 0 & V_1 = 0 & V_1 = 0 \\
V_3 = 0 & V_3 = 0 & V_2 = 0
\end{array}
$$

The column vector of unknowns has the exponential amplitudes at the velocity nodes; these are designated as V_1, V_2 and V_3. The column vector of knowns consists

of exponential amplitudes of the force inputs. The
force amplitude entering node 1 due to the force inputs
is merely F_1. There is no force input entering node 2
directly and thus the zero in the known column vector
at node 2. For node 3 we may observe that force amplitude, F_2, acts away from the node. As a result the force
amplitude is $-F_2$.

We may now determine the admittance matrix. Element
Y_{11} represents the admittance from terminal $1(V_1)$ with
all other terminals grounded ($V_2 = V_3 = 0$). This is the
parallel combination of b_1, $m_1 s$, and b_3, so that:

$$Y_{11} = b_1 + b_3 + m_1 s$$

In a similar way, Y_{22} represents the equivalent admittance from terminal $2(V_2)$ with all other terminals
grounded ($V_1 = V_3 = 0$).

$$Y_{22} = b_3 + \frac{k}{s}$$

and Y_{33} applies to terminal $3(V_3)$ with the other terminals grounded ($V_1 = V_2 = 0$).

$$Y_{33} = b_2 + m_2 s + \frac{k}{s}$$

The other admittances are merely the negative of all
common terms. For example, to find Y_{12} and Y_{21} there
is only one common term, b_3, in Y_{11} and Y_{22}. Thus
$Y_{12} = Y_{21} = -b_3$.

To find Y_{23} and Y_{32} the only common term in Y_{22} and
Y_{33} is k/s and thus $Y_{23} = Y_{32} = -k/s$.

Since there are no common terms between Y_{11} and Y_{33}
the admittances $Y_{13} = Y_{31} = 0$. The complete matrix
is therefore:

$$\begin{bmatrix} b_1 + b_3 + m_1 s & -b_3 & 0 \\ -b_3 & b_3 + k/s & -k/s \\ 0 & -k/s & b_2 + m_2 s + k/s \end{bmatrix}\begin{bmatrix} V_1 \\ V_2 \\ V_3 \end{bmatrix} = \begin{bmatrix} F_1 \\ 0 \\ -F_2 \end{bmatrix}$$

To express the amplitude at any terminal in terms of the basic components and the known input amplitudes we must use determinants. Thus, for example:

$$V_1 = \frac{\begin{vmatrix} F_1 & -b_3 & 0 \\ 0 & b_3 + k/s & -k/s \\ -F_2 & -k/s & b_2 + m_2 s + k/s \end{vmatrix}}{\begin{vmatrix} b_1 + b_3 + m_1 s & -b_3 & 0 \\ -b_3 & b_3 + k/s & -k/s \\ 0 & -k/s & b_2 + m_2 s + k/s \end{vmatrix}}$$

or if we evaluate the determinants

$$V_1 = \frac{F_1[(b_3+k/s)(D)-(-k/s)(-k/s)]+(-F_2)[(-b_3)(-k/s)]}{(b_1+b_3+m_1 s)[(b_3+k/s)(D)-(-k/s)(-k/s)]-[(-b_3)^2(D)]}$$

where $D = (b_2 + m_2 s + k/s)$.

We may find expressions for V_2 and V_3 in a similar way.

Example 7.9

Find the node-to-datum equations in matrix form for the electrical circuit shown in Figure 7.16a. Also express the amplitude of e_2 in terms of the components and input sources.

Figure 7.16b shows the admittance diagram representation of the electrical circuit. There is a total of four nodes or terminals. However, two of them (the ground terminal and the voltage source terminal, E) are

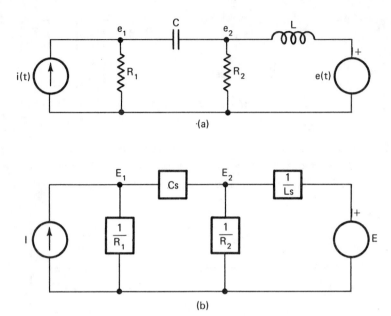

Figure 7.16 Electrical circuit (example 7.9)

already defined. Thus, there are only two independent
nodes and the admittance matrix will be of the form:

$$
\begin{bmatrix} Y_{11} & Y_{12} \\ \\ Y_{21} & Y_{22} \end{bmatrix}
\begin{bmatrix} E_1 \\ \\ E_2 \end{bmatrix}
=
\begin{bmatrix} I \\ \\ E/Ls \end{bmatrix}
\quad
\begin{matrix} \leftarrow & \text{node } 1 \\ \\ \leftarrow & \text{node } 2 \end{matrix}
$$

with, above E_1 and E_2, the label "unknowns"; above I and E/Ls, the label "knowns"; and below the matrix, $E_2=0$ under column 1 and $E_1=0$ under column 2.

   First let us consider the flow into the nodes from
the inputs. The flow into node 1 from the input sources
is merely the amplitude of the current source, I. However, in this case the flow into node 2 depends on the
voltage source, E and the inductive admittance, 1/Ls.
This is a flow, E/Ls. As a result the known column
vector is as indicated above.

Matrix admittance element, Y_{11} is the parallel combination of $1/R_1$ and Cs. This is the condition that exists when all terminals except E_1 are grounded. Thus:

$$Y_{11} = \frac{1}{R_1} + Cs$$

To find matrix element Y_{22} we reverse the procedure. Now all terminals except E_2 are grounded. The admittance from node 2 in this case is the parallel combination of C, $1/R_2$, and $1/Ls$ or

$$Y_{22} , = \frac{1}{R_2} + \frac{1}{Ls} + Cs$$

In this case there is only one common term between Y_{11} and Y_{12}, and it is Cs, so that:

$$Y_{12} = Y_{21} = - Cs$$

and the complete matrix representation is:

$$\begin{bmatrix} \frac{1}{R_1} + Cs & -Cs \\ -Cs & \frac{1}{R_2} + \frac{1}{Ls} + Cs \end{bmatrix} \begin{bmatrix} E_1 \\ E_2 \end{bmatrix} = \begin{bmatrix} I \\ E/Ls \end{bmatrix}$$

The amplitude of E_2 in terms of determinants is:

$$E_2 = \frac{\begin{vmatrix} \frac{1}{R_1} + Cs & I \\ -Cs & E/Ls \end{vmatrix}}{\begin{vmatrix} \frac{1}{R_1} + Cs & -Cs \\ -Cs & \frac{1}{R_2} + \frac{1}{Ls} + Cs \end{vmatrix}}$$

and evaluating the determinants

$$E_2 = \frac{(\frac{1}{R_1} + Cs)(\frac{E}{Ls}) - (-Cs)(I)}{(\frac{1}{R_1} + Cs)(\frac{1}{R_2} + \frac{1}{Ls} + Cs) - (-Cs)(-Cs)}$$

Loop and Mesh Equations

The loop and mesh equations method is a convenient means of formulating equations when the unknowns are of the through variable type. This method favors the use of impedance parameters. Let us consider, therefore, the circuit shown in Figure 7.17. The circuit represents a particular fluid circuit that has input pressure sources of exponential type with amplitudes P_1 and P_2. We want to find equations that describe the exponential flows through all the impedance elements. In this case there are three distinct impedances and three distinct flows. However, only two of these are independent. For example, if Q_1 is the flow through Z_b and Q_2 is the flow through Z_c, then the flow through Z_a is $Q_1 - Q_2$. The flows Q_1 and Q_2 are the mesh flows and are the most convenient choice for the unknowns since we may apply Equation (7.38b) around the closed mesh. If we apply Equation (7.38b) (KVL) around meshes 1 and 2:

$$\text{mesh } 1 \rightarrow (P_1 - P_a) + (P_a - P_g) = (P_1 - P_g) \quad (7.44a)$$

$$\text{mesh } 2 \rightarrow (P_g - P_a) + (P_a - P_2) = (P_g - P_2) \quad (7.44b)$$

The meshes are the open spaces in the impedance circuit. We always take the flow in each mesh in a clockwise direction. We may now substitute the component relations into the mesh Equations (7.44a) and (7.44b) to obtain:

$$\text{mesh } 1 \rightarrow Z_b Q_1 + Z_a(Q_1 - Q_2) = P_1 \quad (7.45a)$$

$$\text{mesh } 2 \rightarrow Z_a(Q_2 - Q_1) + Z_c Q_2 = -P_2 \quad (7.45b)$$

or if we group the flows together in Equation (7.45) the result is:

$$\text{mesh } 1 \rightarrow (Z_a + Z_b)Q_1 - Z_a Q_2 = P_1 \qquad (7.46a)$$

$$\text{mesh } 2 \rightarrow - Z_a Q_1 + (Z_a + Z_c)Q_2 = - P_2 \qquad (7.46b)$$

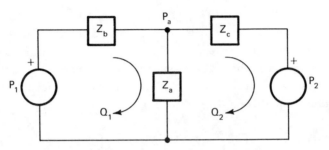

Figure 7.17 Analogous circuit involving impedances

We may interpret the terms in Equation (7.46) as follows:

In Mesh 1: $(Z_a + Z_b)Q_1$ = sum of the pressure drops across each impedance in direction of flow and due to Q_1 only. All other flows set to zero.

$- Z_a Q_2$ = sum of the pressure drops across impedances, in direction Q_1 but due to flow Q_2. All other flows set to zero.

In Mesh 2: $- Z_a Q_1$ = sum of the pressure drops in mesh 2 flow direction due to flow Q_1. All other flows set to zero.

$(Z_a + Z_c)Q_2$ = sum of pressure drops due to flow, Q_2. All other flows set to zero.

In matrix form Equation (7.46) may be written as:

$$
\begin{array}{c} \downarrow \\ \text{unknowns} \end{array} \quad \begin{array}{c} \downarrow \\ \text{knowns} \end{array}
$$

$$
\begin{array}{c} \text{mesh } 1 \rightarrow \\ \\ \text{mesh } 2 \rightarrow \end{array}
\begin{bmatrix} Z_a + Z_b & -Z_a \\ \\ -Z_a & Z_a + Z_c \end{bmatrix}
\begin{bmatrix} Q_1 \\ \\ Q_2 \end{bmatrix}
=
\begin{bmatrix} P_1 \\ \\ -P_2 \end{bmatrix}
\qquad (7.47)
$$

$$
\begin{array}{cc} \uparrow & \uparrow \\ Q_2 = 0 & Q_1 = 0 \end{array}
$$

In the impedance array the terms on the main diagonal are positive and all others are negative. This is true for mesh equations when the assumed flow directions are all clockwise or all counter-clockwise. The general form for the mesh equations is:

$$
\begin{bmatrix} Z_{11} & Z_{12} \\ \\ Z_{21} & Z_{22} \end{bmatrix}
\begin{bmatrix} Q_1 \\ \\ Q_2 \end{bmatrix}
=
\begin{bmatrix} P_1 \\ \\ P_2 \end{bmatrix}
\qquad (7.48)
$$

The element Z_{11} is the sum of all impedances in mesh 1 and Z_{22} is the sum of all impedances in mesh 2. The impedances Z_{12} and Z_{21} are equal and are the negative of all the common terms in Z_{11} and Z_{22} provided all the circulating flows are taken in the same direction (e.g. all clockwise).

We may also now formulate the equilibrium equations for flow variables by the mesh method and by inspection of the impedance circuit. We will demonstrate the procedure in the following examples.

Example 7.10

Figure 7.18 shows an electrical circuit that consists of two resistances, a capacitance, an inductance and two sources. In Figure 7.18a the current source is on the left and the voltage source on the right. The same circuit with sources reversed in position is shown in Figure 7.18b. In each case, find the relationship

between the current through the capacitance and the
sources.

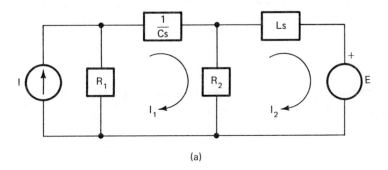

(a)

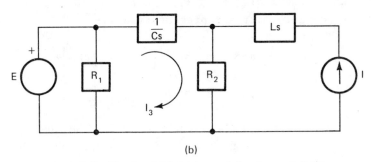

(b)

Figure 7.18 Circuits with impedances shown (example 7.10)

There are three meshes in the circuit shown in Figure
7.18a. However, the current is prescribed in the leftmost mesh. Thus, there are only two unknown currents
and the mesh matrix has the form:

$$
\begin{array}{c}
\text{mesh } 1 \rightarrow \\[20pt]
\text{mesh } 2 \rightarrow
\end{array}
\begin{bmatrix}
Z_{11} & Z_{12} \\
Z_{21} & Z_{22}
\end{bmatrix}
\begin{bmatrix}
I_1 \\
I_2
\end{bmatrix}
=
\begin{bmatrix}
R_1 I \\
-E
\end{bmatrix}
$$

where the right-hand side shows the effect of the
sources. In mesh 1 (I_1) the current source (I) contributes a voltage rise, $R_1 I$. In mesh 2 (I_2) the voltage

source (E) causes a voltage drop and thus is described as -E.

Now let us consider the impedance matrix. The term Z_{11} represents the impedance around mesh 1 and is:

$$Z_{11} = R_1 + \frac{1}{Cs} + R_2$$

Similarly the term Z_{22} is the total impedance around mesh 2. This is:

$$Z_{22} = R_2 + Ls$$

There is only one common factor between Z_{11} and Z_{22}, and that is the resistance, R_2. Thus $Z_{12} = Z_{21} = -R_2$ and the complete matrix representation is:

$$\begin{bmatrix} R_1 + \frac{1}{Cs} + R_2 & -R_2 \\ -R_2 & R_2 + Ls \end{bmatrix} \begin{bmatrix} I_1 \\ I_2 \end{bmatrix} = \begin{bmatrix} R_1 I \\ -E \end{bmatrix}$$

The current through the capacitor is I_1. We therefore solve the equations for I_1. The solution has the form of the ratio of two determinants.

$$I_1 = \frac{\begin{vmatrix} R_1 I & -R_2 \\ -E & R_2 + Ls \end{vmatrix}}{\begin{vmatrix} R_1 + \frac{1}{Cs} + R_2 & -R_2 \\ -R_2 & R_2 + Ls \end{vmatrix}}$$

$$= \frac{R_1 I (R_2 + Ls) - (-E)(-R_2)}{(R_1 + \frac{1}{Cs} + R_2)(R_2 + Ls) - (-R_2)(-R_2)}$$

Now let us turn our attention to Figure 7.18b. Once again, there are three meshes. In this case two of them are special cases. The only unknown current is through the capacitance (I_3). Furthermore, the components R_1 and L have no effect on the unknown current. That is, the contribution to mesh (I_3) from the leftmost mesh is always a voltage rise, E, regardless of the magnitude of R_1. The contribution to mesh (I_3) from the rightmost mesh is always a voltage drop, $R_2 I$, and it is independent of L. The matrix thus degenerates to:

$$Z_{33} \; I_3 = E - R_2 I$$

and

$$Z_{33} = R_2 + \frac{1}{Cs}$$

so that

$$I_3 = \frac{E - R_2 I}{R_2 + \frac{1}{Cs}}$$

We may observe that the current through the capacitor is vastly different when the sources are reversed.

Example 7.11

Figure 7.19a shows a mechanical circuit with two dampers and a mass. This mechanical circuit may be represented by either of the two impedance circuits shown in Figure 7.19b and c. Show that the force acting to accelerate the mass is the same in both circuits.

In circuit 7.19b the mesh equations matrix has the form:

$$\begin{bmatrix} Z_{11} & Z_{12} \\ Z_{21} & Z_{22} \end{bmatrix} \begin{bmatrix} F_1 \\ F_2 \end{bmatrix} = \begin{bmatrix} V \\ 0 \end{bmatrix}$$

where the source has no contribution in mesh (F_2) and contributes a velocity rise (V) to mesh (F_1). The sum of the impedances around the meshes yields:

$$Z_{11} = \frac{1}{b_1} + \frac{1}{ms} \ , \quad Z_{22} = \frac{1}{b_2} + \frac{1}{ms}$$

and the common term is $1/ms$ so that $Z_{12} = Z_{21} = -1/ms$ and therefore the matrix becomes:

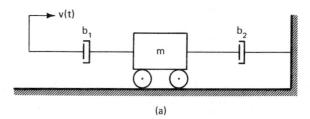

$$\begin{bmatrix} \dfrac{1}{b_1} + \dfrac{1}{ms} & -\dfrac{1}{ms} \\[2mm] -\dfrac{1}{ms} & \dfrac{1}{b_2} + \dfrac{1}{ms} \end{bmatrix} \begin{bmatrix} F_1 \\[2mm] F_2 \end{bmatrix} = \begin{bmatrix} V \\[2mm] 0 \end{bmatrix}$$

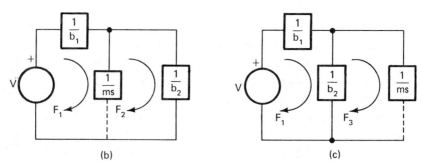

(a)

(b)   (c)

Figure 7.19 Mechanical circuit (example 7.11)

In this circuit (Figure 7.19b) the force applied to the mass is $F_M = F_1 - F_2$. We must therefore solve the matrix for F_1 and F_2 and then subtract the results.

$$F_1 = \frac{\begin{vmatrix} V & -\dfrac{1}{ms} \\[2ex] 0 & \dfrac{1}{b_2} + \dfrac{1}{ms} \end{vmatrix}}{\begin{vmatrix} \dfrac{1}{b_1} + \dfrac{1}{ms} & -\dfrac{1}{ms} \\[2ex] -\dfrac{1}{ms} & \dfrac{1}{b_2} + \dfrac{1}{ms} \end{vmatrix}}$$

$$= \frac{V\left(\dfrac{1}{b_2} + \dfrac{1}{ms}\right)}{\left(\dfrac{1}{b_1} + \dfrac{1}{ms}\right)\left(\dfrac{1}{b_2} + \dfrac{1}{ms}\right) - \dfrac{1}{m^2 s^2}}$$

$$F_2 = \frac{\begin{vmatrix} \dfrac{1}{b_1} + \dfrac{1}{ms} & V \\[2ex] -\dfrac{1}{ms} & 0 \end{vmatrix}}{\begin{vmatrix} \dfrac{1}{b_1} + \dfrac{1}{ms} & -\dfrac{1}{ms} \\[2ex] -\dfrac{1}{ms} & \dfrac{1}{b_2} + \dfrac{1}{ms} \end{vmatrix}}$$

$$= \frac{\dfrac{V}{ms}}{\left(\dfrac{1}{b_1} + \dfrac{1}{ms}\right)\left(\dfrac{1}{b_2} + \dfrac{1}{ms}\right) - \dfrac{1}{m^2 s^2}}$$

and thus:

$$F_M = F_1 - F_2 = \frac{\dfrac{V}{b_2}}{\left(\dfrac{1}{b_1} + \dfrac{1}{ms}\right)\left(\dfrac{1}{b_2} + \dfrac{1}{ms}\right) - \dfrac{1}{m^2 s^2}}$$

$$= \frac{\dfrac{V}{b_2}}{\dfrac{1}{b_1 b_2} + \dfrac{1}{ms}\left(\dfrac{1}{b_1} + \dfrac{1}{b_2}\right)}$$

Let us solve the same problem by using the circuit shown in Figure 7.19c. The mesh equations by inspection are:

$$\begin{bmatrix} \dfrac{1}{b_1} + \dfrac{1}{b_2} & -\dfrac{1}{b_2} \\[2em] -\dfrac{1}{b_2} & \dfrac{1}{b_2} + \dfrac{1}{ms} \end{bmatrix} \begin{bmatrix} F_1 \\[2em] F_3 \end{bmatrix} = \begin{bmatrix} V \\[2em] 0 \end{bmatrix}$$

Note that this matrix is different from the matrix determined for Figure 7.19b. In this circuit the force
applied to the mass is F_3 alone. Thus if we solve for
F_3 we obtain:

$$F_M = F_3 = \frac{\begin{vmatrix} \dfrac{1}{b_1} + \dfrac{1}{b_2} & V \\[2em] -\dfrac{1}{b_2} & 0 \end{vmatrix}}{\begin{vmatrix} \dfrac{1}{b_1} + \dfrac{1}{b_2} & -\dfrac{1}{b_2} \\[2em] -\dfrac{1}{b_2} & \dfrac{1}{b_2} + \dfrac{1}{ms} \end{vmatrix}}$$

$$= \frac{\dfrac{V}{b_2}}{\left(\dfrac{1}{b_1} + \dfrac{1}{b_2}\right)\left(\dfrac{1}{b_2} + \dfrac{1}{ms}\right) - \dfrac{1}{b_2^2}}$$

or

$$F_M = \frac{\dfrac{V}{b_2}}{\dfrac{1}{b_1 b_2} + \dfrac{1}{ms}\left(\dfrac{1}{b_1} + \dfrac{1}{b_2}\right)}$$

and the results are the same in both circuits.

7.6 SOME PROPERTIES OF LINEAR SYSTEMS

Much of the simplification of the process of writing
equilibrium equations in Section 7.5 is due to the basic
property of the elements of a linear system: responses
are directly proportional to the driving function (or
its derivative or integral). Indeed, this is why the
equivalent impedance of a series connection is simply
the sum of the individual impedances. It should therefore be expected that these basic attributes of the

elements which make up a system should result in
characteristic properties of a linear system. Some of
these properties are expressed as Circuit Theorems which
greatly simplify the task of eliminating unwanted variables.
Others are useful in that they give guidance as to the
kinds of measurements which must be taken to obtain a
valid model of a system.

Consider first a set of node equations, the basic form
of which we have seen is:

$$[Y][V] = [F] \tag{7.49}$$

In the case of a circuit having three independent nodes
the details are:

$$\begin{bmatrix} Y_{11} & Y_{12} & Y_{13} \\ Y_{21} & Y_{22} & Y_{23} \\ Y_{31} & Y_{32} & Y_{33} \end{bmatrix} \begin{bmatrix} V_1 \\ V_2 \\ V_3 \end{bmatrix} = \begin{bmatrix} F_1 \\ F_2 \\ F_3 \end{bmatrix} \tag{7.50}$$

The solution has the basic form:

$$[V] = [Y]^{-1}[F] = [z][F] \tag{7.51}$$

where

$$[z] = \begin{bmatrix} z_{11} & z_{12} & z_{13} \\ z_{21} & z_{22} & z_{23} \\ z_{31} & z_{32} & z_{33} \end{bmatrix}$$

$$= \frac{1}{\Delta} \begin{bmatrix} \Delta_{11} & -\Delta_{21} & \Delta_{31} \\ -\Delta_{12} & \Delta_{22} & -\Delta_{32} \\ \Delta_{31} & -\Delta_{32} & \Delta_{33} \end{bmatrix}$$

$$\Delta = \text{the determinant of } [Y]$$

and

$$\Delta_{jk} = \text{the co-factor of } y_{jk}$$

The terms in the inverse matrix have some physical significance, in that they are driving point and transfer impedances. That is, the term z_{jj} is the impedance between node j and ground when the sources connected to all other nodes are suppressed (i.e., set to zero), which can be recognized as the driving point impedance discussed in Section 7.3. The term z_{jk} is the ratio of the across variable between node j and ground to the through variable entering node k when all other sources are suppressed, usually called the transfer impedance.

It is therefore theoretically possible to take appropriate measurements on a system to yield values for the entire array of driving point and transfer impedances so that the matrix [z] may be measured. By inverting it the matrix [Y] of the corresponding node equations can then be obtained. Reversal of the procedure to obtain the node equations by inspection of the circuit will then produce a circuit model of the system.

In practice, it is rarely necessary to carry out this inversion process because the form of the circuit model can be determined, knowing the number of variables required. However, this does indicate the form of the measurements which must be taken on a system, including the necessary constraints to be applied.

It should be evident that if we consider a set of mesh equations having the form:

$$[Z][F] = [V] \qquad\qquad (7.52)$$

the solution is:

$$[F] = [Z]^{-1}[V] = [y][V] \qquad\qquad (7.53)$$

where [y] is the array of driving point and transfer admittances. Note that there is no simple relation between
[y] (i.e., the driving point and transfer admittances) and
[Y] (i.e., the admittance array of the node equations).
Nor is there any simple relation between the driving point
and transfer impedance matrix [z] and the mesh equation
impedance matrix [Z].

Superposition Theorem

Returning to Equation (7.51) we can write out the
solution for V_1 as:

$$V_1 = z_{11} F_1 + z_{12} F_2 + z_{13} F_3 \qquad (7.54)$$

We note that for this system having three sources, the
solution has three components each of which is the value
of V_1 while the other sources are set to zero. The Superposition Theorem uses this property of a linear system to
obtain the total response as the sum of the responses to
each driving function or source taken one at a time while
all other sources are suppressed.

Example 7.12

A circuit consisting of two resistors is driven by a
voltage source and a current source as shown in Figure
7.20a. Find the voltage across R_2.

The procedure is simply to obtain the values of E_2
when only one source is acting at a time, and to add
those solutions. In this case there are two sources,
so the problem is to find E_2 in each of Figures 7.20b
and 7.20c and to add them.

From Figure 7.20b the first component of E_2 can be
found directly by noting that this is a voltage divider.

$$E_2 = \frac{R_2}{R_1 + R_2} \cdot E$$

From Figure 7.20c the second component of E_2 is found
by noting that this is the voltage across the parallel
combination of R_1 and R_2.

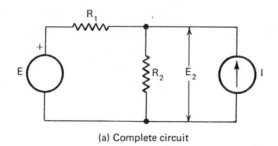

(a) Complete circuit

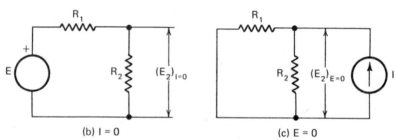

(b) I = 0   (c) E = 0

Figure 7.20 Superposition

$$E_2 = \frac{R_1 R_2}{R_1 + R_2} \cdot I$$

The complete response is the sum of those two expressions:

$$E_2 = \frac{R_2 E}{R_1 + R_2} + \frac{R_1 R_2 I}{R_1 + R_2}$$

$$= \frac{R_2}{R_1 + R_2} [E + R_1 I]$$

Reciprocity Theorem

We have previously noted that when all of the elements in a system or circuit are bilateral the terms Y_{jk} and Y_{kj} are equal in the node equations and Z_{jk} and Z_{kj} are equal in the mesh equations, and therefore the matrices [Y] and [Z] are symmetrical. As a result, their inverses, [z] and [y], are also symmetrical.

Although this is of little value as an aid to analyze a circuit, a corollary of the theorem is of some value in modelling. If a set of measurements on a system shows that any pair of transfer impedances or admittances is not reciprocal (i.e., $z_{jk} \neq z_{kj}$ or $y_{jk} \neq y_{kj}$) the system cannot be modelled using only bilateral elements and independent sources such as those being considered in this text.

Thevenin's Theorem

Thevenin's Theorem is a very convenient method of obtaining a simple circuit which is equivalent to a section of a complete circuit. To some extent, it is an extension of the modelling of real sources as an ideal source associated with an impedance.

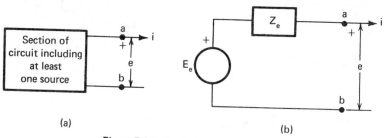

(a)   (b)

Figure 7.21 Equivalent circuit

The section of the circuit to be modelled or simplified will always have at least one source and a pair of terminals at which it is connected to the remainder of the complete circuit. This is shown symbolically in Figure 7.21a. The theorem states simply that the circuit shown in Figure 7.21b will produce exactly the same current-voltage relationships at terminals a and b provided:

(a) E_e is the voltage between a and b when $i = 0$ (i.e., the open circuit voltage).

(b) Z_e is the impedance between terminals a and b when all the sources within the section of the circuit are suppressed (i.e., the driving point impedance).

Example 7.13

Obtain the system function E_L/E for the circuit shown in Figure 7.22a.

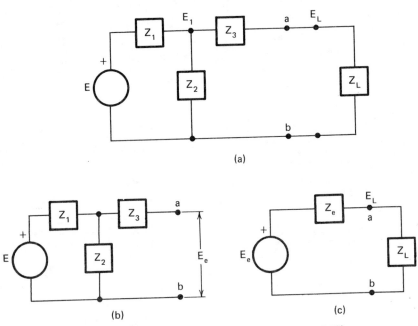

(a)

(b)   (c)

Figure 7.22 Thevenin equivalent circuit (example 7.13)

The procedure is to obtain the Thevenin equivalent of the section of the original circuit reproduced in Figure 7.22b. Since there is no current flow through Z_3 the open circuit voltage is obtained directly from the voltage divider relation:

$$E_e = \frac{Z_2}{Z_1 + Z_2} \cdot E$$

The equivalent impedance is the driving point impedance between terminals a and b when the source is suppressed and this is readily seen to be the equivalent of Z_3 in series with the parallel combination of Z_1 and Z_2.

$$Z_e = Z_3 + \frac{Z_1 Z_2}{Z_1 + Z_2}$$

$$= \frac{Z_1 Z_2 + Z_2 Z_3 + Z_3 Z_1}{Z_1 + Z_2}$$

Using these values in Figure 7.22c we obtain:

$$\frac{E_L}{E_e} = \frac{Z_L}{Z_e + Z_L}$$

and substituting for E_e and Z_e gives

$$\frac{E_L}{E} = \frac{Z_2}{Z_1 + Z_2} \cdot \frac{Z_L}{Z_e + Z_L}$$

$$= \frac{Z_2 Z_L}{Z_1 Z_2 + Z_2 Z_3 + Z_3 Z_1 + Z_L (Z_1 + Z_2)}$$

This circuit is sufficiently simple to check this answer by obtaining the solution to the node equations of the entire circuit.

$$\begin{bmatrix} \frac{1}{Z_1} + \frac{1}{Z_2} + \frac{1}{Z_3} & -\frac{1}{Z_3} \\ -\frac{1}{Z_3} & \frac{1}{Z_3} + \frac{1}{Z_L} \end{bmatrix} \begin{bmatrix} E_1 \\ E_L \end{bmatrix} = \begin{bmatrix} \frac{E}{Z_1} \\ 0 \end{bmatrix}$$

The solution is

$$E_L = \frac{\dfrac{E}{Z_3 Z_1}}{(\frac{1}{Z_1} + \frac{1}{Z_2} + \frac{1}{Z_3})(\frac{1}{Z_3} + \frac{1}{Z_L}) - \frac{1}{Z_3^2}}$$

$$= \frac{Z_2 E}{\frac{1}{Z_L}(Z_1 Z_2 + Z_2 Z_3 + Z_3 Z_1) + Z_1 + Z_2}$$

$$= \frac{Z_2 Z_L E}{Z_1 Z_2 + Z_2 Z_3 + Z_3 Z_1 + Z_L (Z_1 + Z_2)}$$

In some cases the process of obtaining the Thevenin equivalent by the simple procedures used in Example 7.13 becomes rather complicated and it is simpler to obtain the open circuit voltage and driving point impedance directly from a set of node equations. This process is shown in the following example.

Example 7.14

Find an expression for the current through the resistor R in the circuit shown in Figure 7.23a.

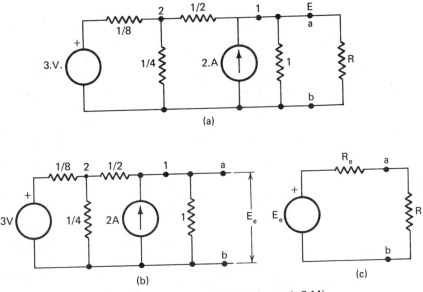

Figure 7.23 Electrical circuit (example 7.14)

In this case the section of the circuit for which the Thevenin equivalent is required consists of all elements and sources other than R. For clarity this is shown in Figure 7.23b, but normally should not be necessary. The node equations are:

$$\begin{bmatrix} 3 & -2 \\ -2 & 14 \end{bmatrix} \begin{bmatrix} E_1 \\ E_2 \end{bmatrix} = \begin{bmatrix} 2 \\ 24 \end{bmatrix}$$

The open circuit voltage, E_e, for this circuit is:

$$E_1 = \frac{1}{\Delta} \begin{vmatrix} 2 & -2 \\ 24 & 14 \end{vmatrix}$$

where

$$\Delta = \begin{vmatrix} 3 & -2 \\ -2 & 14 \end{vmatrix} = 38$$

That is:

$$E_e = E_1 = 2 \text{ V}$$

The equivalent impedance is the driving point impedance between node 1 and ground. The simplest way to obtain this is the replacement of the actual sources by a vector having only one source injecting the current I at node 1. That is the equations are now:

$$\begin{bmatrix} 3 & -2 \\ -2 & 14 \end{bmatrix} \begin{bmatrix} E_1 \\ E_2 \end{bmatrix} = \begin{bmatrix} I \\ 0 \end{bmatrix}$$

The required driving point impedance is the ratio:

$$\frac{E_1}{I} = z_{11} = \frac{14}{38} = \frac{7}{19} \text{ ohm}$$

Thus the Thevenin equivalent shown in Figure 7.23c has $E_e = 2$ V and $R_e = 7/19$ ohm, and hence the current through resistor R is given by:

$$I = \frac{2}{R + \frac{7}{19}}$$

A simple application of Thevenin's Theorem which is often very useful is that of transforming a through vari-

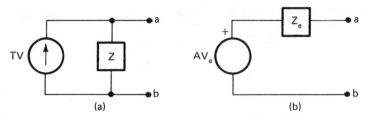

Figure 7.24 Application of Thevenin's Theorem. (a) Original
circuit; (b) Thevenin equivalent circuit

able source and the impedance connected across its terminals (Figure 7.24a) to an equivalent series connected
across variable source and impedance (Figure 7.24b). The
circuits are equivalent provided $(AV)_e = Z(TV)$ and $Z_e = Z$.

Norton's Theorem

This is the dual of Thevenin's Theorem in that the
equivalence for a section of a circuit is expressed in
terms of a through variable source and shunting admittance.
This is indicated in Figure 7.25 where 7.25b is the
equivalent of 7.25a provided:
  (a) I_e is the current flow when the voltage between a
      and b is zero (i.e., the short-circuit current).
  (b) Y_e is the admittance between terminals a and b when
      all the sources within the section of the circuit
      are suppressed (i.e., the driving point admittance).

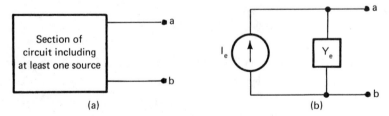

Figure 7.25 Norton equivalent circuit

The Norton equivalent can be obtained by series-parallel
combinations, or by solution of mesh equations, or by a
transformation of the Thevenin equivalent.

In all the equivalent circuits we have considered it is important to note that some information is lost when working with the equivalent. The equivalence is valid only at the terminals we have designated a and b. In effect, these equivalences are convenient methods to eliminate unwanted variables, thereby simplifying some of the mathematical manipulation which is required.

Problems

7.1 Find the driving point impedance, Z_{dp}, for the circuit diagram shown in Figures P7.1a and P7.1b.

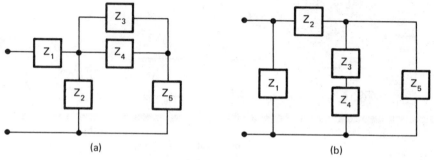

(a)   (b)

Figure P 7.1

7.2 Find the driving point admittance, Y_{dp}, for Figures P7.2a and P7.2b.

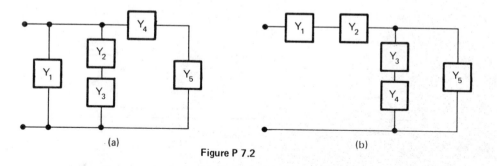

(a)   (b)

Figure P 7.2

7.3 Determine the driving point impedance, $Z_{dp}(s)$, and the driving point admittance, $Y_{dp}(s)$ for the electric circuits shown in Figures P7.3a and P7.3b.

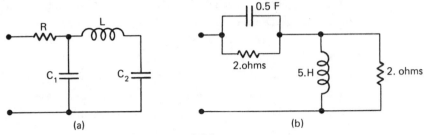

(a)   (b)

Figure P 7.3

7.4 Derive the driving point impedance, Z(s), (from
terminal 1 to g) for the mechanical systems given
in Figures P7.4a and P7.4b.

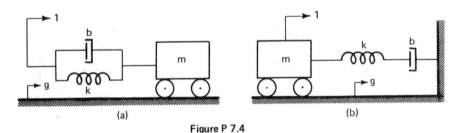

(a)   (b)

Figure P 7.4

7.5 A fluid system is shown schematically in Figure P7.5.
The section at the left end of R_1 is designated as
terminal 1. Find the driving point admittance (from
terminal 1 to g) as a function of s.

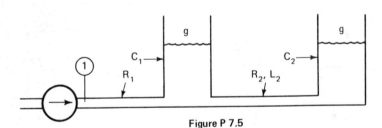

Figure P 7.5

7.6 Derive the system function (output/input) for the
circuits shown in Figures P7.6a and P7.6b.

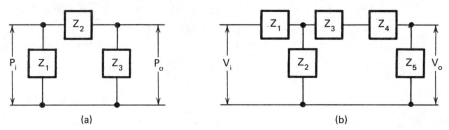

(a) (b)

Figure P 7.6

7.7 Find the system function ($G(s) = E_o/E_i$) for the electric circuits given in Figures P7.7a and P7.7b.

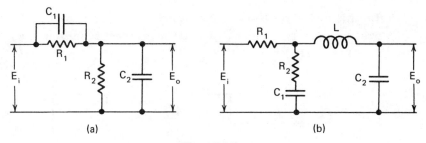

(a) (b)

Figure P 7.7

7.8 A torque (T_i) applied to flywheel 1, is transmitted to flywheel 2 through a shaft that may be modelled as a torsional spring and damper in parallel. The torque developed in flywheel 2 is designated as T_o (Figure P7.8). Find the system function, ($G(s) = T_o/T_i$).

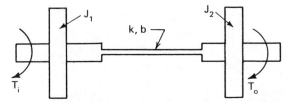

Figure P 7.8

7.9 A force $f(t) = F e^{st}$ is applied to mass 1 in Figure
     P7.9. The velocity of mass 3, is assumed to have
     the form $v_3(t) = V_3 e^{st}$. Find the system function
     (V_3/F) in terms of m_1, m_2, m_3, b, k, and s.

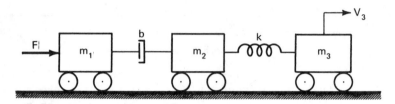

Figure P 7.9

7.10 In the fluid system (Figure P7.10) an air pump
     pressurizes two tanks. Find the system function
     $(G(s) = P_0/P_i)$.

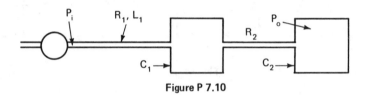

Figure P 7.10

7.11 The input and output voltages are indicated on the
     electric circuit in Figure P7.11. Find the system
     function, E_0/E_i.

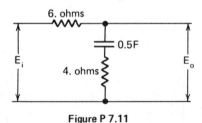

Figure P 7.11

7.12 The input to a mechanical system (Figure P7.12) is
     a velocity of the form $v(t) = V e^{st}$. The output
     is designated as the force transmitted through

damper, b_2 and it is assumed to be $f_b(t) = F\,e^{st}$. Determine the system function with these signals as output and input.

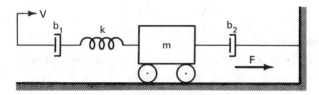

Figure P 7.12

7.13 A constant flow pump supplies water to a reservoir with a drain in it. The system can be modelled as indicated in Figure P7.13. Find the system function that describes the ratio of output flow to input flow.

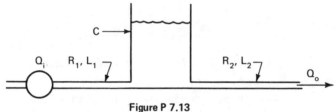

Figure P 7.13

7.14 A mechanical system has a system function that has been derived as:

$$\frac{V_o}{V_i} = \frac{s}{s + 10}$$

Find $v_o(t)$ if $v_i(t) = 10\,u_s(t) + 5\,u_s(t - 2)$.

7.15 The system function for a particular fluid system is:

$$\frac{P_o}{P_i} = \frac{s + 1}{s + 2}$$

If $p_i(t) = 3\,u_r(t)$, find $p_o(t)$.

7.16 The output flow is related to the input pressure in a fluid circuit by the system function

$$\frac{Q_o}{P_i} = \frac{s + 4}{s^2 + 3s + 2}$$

Determine $q_o(t)$ if $p_i(t) = u_s(t)$.

7.17 The ratio of output force to input velocity for a mechanical system is given as:

$$\frac{F_o}{V_i} = \frac{1}{s^2 + 12s + 100}$$

The input velocity may be described by $v_i(t) = 2\, u_i(t)$. What is the output force as a function of time?

7.18 An electric circuit has the system function:

$$\frac{E_o}{E_i} = \frac{s}{s^2 + 4s + 4}$$

An input $e_i(t) = 5\, u_s(t)$ is applied. What is the output voltage, $e_o(t)$?

7.19 Write a set of node equations in matrix form for the physical systems shown in Figures P7.19a, P7.19b, P7.19c and P7.19d.

7.20 Write a set of mesh equations in matrix form for the physical circuits indicated in Figures P7.20a, P7.20b, P7.20c and P7.20d.

7.21 For the network shown in Figure P7.21 find the output voltage, $e_o(t)$, if the input voltage, $e_i(t) = u_s(t)$.

Hint: Use node equations to find system function E_o/E_i and then derive differential equation.

7.22 For the resistive network shown in Figure P7.22 the source voltage $e_s = 100$ V. Using a Thevenin equivalent circuit determine the voltage, e_3, in terms of the unknown resistance, R.

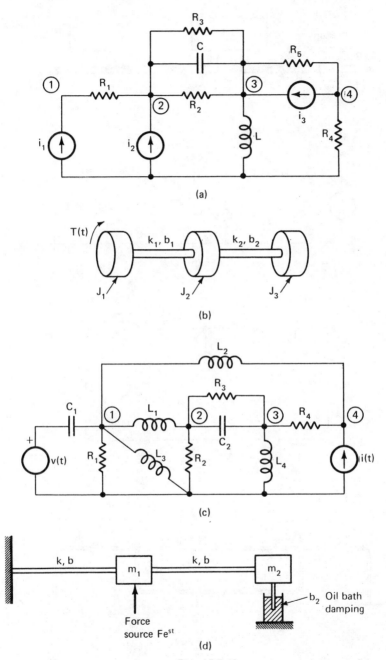

(a)

(b)

(c)

(d)

Figure P 7.19

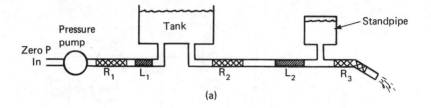

(a)

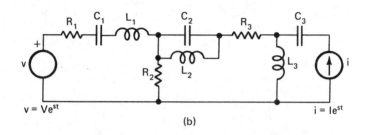

$v = Ve^{st}$ $i = Ie^{st}$

(b)

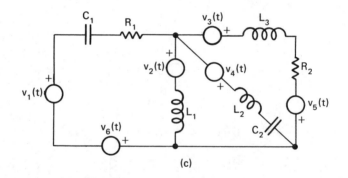

(c)

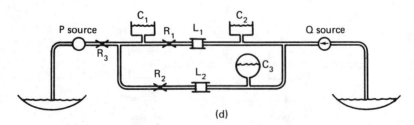

(d)

Figure P 7.20

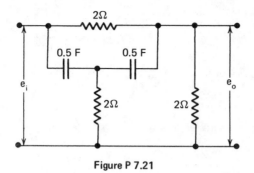

Figure P 7.21

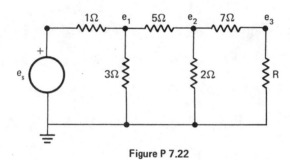

Figure P 7.22

CHAPTER 8

SINUSOIDAL RESPONSE

In Chapter V we noted that the response of a circuit is
the time variation that occurs in one signal variable as
the result of a time variation in another signal variable.
We may also express the idea of response as the output
for a specific input. For example, the step, ramp or
impulse responses are the outputs when the inputs are a
step, ramp, or impulse respectively. The common factor
in all these responses is that the inputs and outputs are
both functions of time.

When a periodic function (e.g. sine or cosine) is applied
as an input to a circuit we might term the output as the
sinusoidal (or cosinusoidal) response. Here again, the
input and output would be functions of time. This would

then be consistent with our previous treatment of other
input functions. In this chapter, we will apply sine and
cosine inputs to physical circuits. Upon examination of
the responses to these inputs we will notice a transient
(non-periodic) portion of the response and a periodic
portion of the response. In some situations, such as
analyzing fault currents in a power system, the complete
response is important and must be determined. However,
for many applications the transient portion is of no
particular interest and therefore it is not evaluated.
The periodic portion is usually more important and is
designated as the "sinusoidal steady-state" response.

In the sinusoidal steady-state when the input to a linear
system has a sine (or cosine) waveform the output also has
a sine (or cosine) waveform at the same frequency. The
sinusoidal steady state response is distinguished by the
amplitude and phase shift of the output waveform relative
to the input waveform. This relative amplitude and phase
shift property of the sinusoidal steady state response is
a function of the waveform frequency. As a result the
amplitude-frequency and the phase shift-frequency relations
completely specify the response. These relations are
referred to as the "frequency response".

We will begin the development of the sinusoidal response
relations by applying direct mathematical methods. Then
we will use impedance and system function concepts to
obtain the same results.

8.1 RESPONSE TO SINE AND COSINE FUNCTIONS

First Order Circuits

From Equation (5.1) the most general differential equation
for first order linear time invariant circuits is:

$$\tau \frac{dy}{dt} + y = a_1 \, g(t) + a_2 \, \frac{dg(t)}{dt} \qquad (8.1)$$

When the input function, $g(t)$, is a sine wave function of
amplitude, G, and frequency, ω, then $g(t) = G \sin\omega t$.

Equation (8.1) may therefore be written as:

$$\tau \frac{dy}{dt} + y = a_1 G \sin\omega t + a_2 \omega G \cos\omega t \qquad (8.2)$$

$$y(0^+) = 0$$

(the initial condition is zero for the sine input if there is no initial storage). The complete solution for y consists of the summation of the homogeneous solution and the particular solution (Equation (5.4)). Thus:

$$y = y_H + y_P$$

We know from Chapter V that there is only one homogeneous solution for first order circuits. That homogeneous solution has the form given in Equation (5.8), $A_1 e^{-t/\tau}$.

To find the particular solution we follow the standard mathematical procedure of undetermined coefficients. This method may be used when the input function has only a finite number of independent derivatives. The method consists of summing a term proportional to the input function with terms proportional to all possible derivatives of the input functions. In this case, since the sine and cosine functions are derivatives of each other we have only two terms in the solution. Thus we assume a particular solution of the form:

$$y_P = A_2 \sin\omega t + A_3 \cos\omega t \qquad (8.3)$$

where the coefficients A_2 and A_3 are to be determined by substituting Equation (8.3) into Equation (8.2). That operation yields:

$$[A_2 - \omega\tau A_3]\sin\omega t + [A_3 + \omega\tau A_2]\cos\omega t$$

$$= a_1 G \sin\omega t + a_2 \omega G \cos\omega t \qquad (8.4)$$

If we equate the coefficients of sine and cosine terms

on each side of Equation (8.4) we obtain the simultaneous
equations:

$$A_2 - \omega\tau A_3 = a_1 G$$

$$A_3 + \omega\tau A_2 = a_2 \omega G \tag{8.5}$$

Solution of Equation (8.5) produces:

$$A_2 = \frac{G[a_1 + a_2\omega^2\tau]}{[1 + \omega^2\tau^2]} \tag{8.6}$$

$$A_3 = \frac{G\omega[a_2 - a_1\tau]}{[1 + \omega^2\tau^2]}$$

Thus the particular solution of Equation (8.3) becomes:

$$y_P = \frac{G}{1 + \omega^2\tau^2}[(a_1+a_2\omega^2\tau)\sin\omega t+(a_2\omega-a_1\omega\tau)\cos\omega t] \tag{8.7}$$

Equation (8.7) may also be expressed in terms of a phase
shifted sine wave as indicated in Chapter II. Since the
input g(t) = G sinωt we would like to represent the forced
output (particular response) as A sin(ωt + φ) where A is
an amplitude and φ is a phase shift. In those terms
Equation (8.7) becomes:

$$y_P = \frac{G(a_1^2+a_2^2\omega^2)^{\frac{1}{2}}}{(1+\omega^2\tau^2)^{\frac{1}{2}}}\sin(\omega t + \phi)$$

where

$$\phi = \tan^{-1}\frac{a_2\omega - a_1\omega\tau}{a_1 + a_2\omega^2\tau} \qquad \sin\phi = \frac{(a_2 - a_1\tau)\omega}{\sqrt{(a_1^2+a_2^2\omega^2)(1+\omega^2\tau^2)}} \tag{8.8}$$

The complete solution for y(t) in Equation (8.2) now
has the form:

$$y(t) = A_1 e^{-t/\tau} + \frac{G(a_1^2+a_2^2\omega^2)^{\frac{1}{2}}}{(1+\omega^2\tau^2)^{\frac{1}{2}}}\sin(\omega t + \phi) \tag{8.9}$$

We may evaluate the constant A_1 in Equation (8.9) by applying the initial condition $y(0^+) = 0$:

$$y(0^+) = A_1 + \frac{G(a_1^2 + a_2^2 \omega^2)^{\frac{1}{2}}}{(1 + \omega^2 \tau^2)^{\frac{1}{2}}} \sin\phi = 0 \qquad (8.10)$$

from which we obtain:

$$A_1 = \frac{-G(a_1^2 + a_2^2 \omega^2)^{\frac{1}{2}} \sin\phi}{(1 + \omega^2 \tau^2)^{\frac{1}{2}}} = \frac{G(a_1 \omega\tau - a_2 \omega)}{(1 + \omega^2 \tau^2)} \qquad (8.11)$$

As a result the complete output response for a sine wave input is:

$$y(t) = \frac{G(a_1 \omega\tau - a_2 \omega)}{1 + \omega^2 \tau^2} e^{-t/\tau} + \frac{G(a_1^2 + a_2^2 \omega^2)^{\frac{1}{2}}}{(1 + \omega^2 \tau^2)^{\frac{1}{2}}} \sin(\omega t + \phi) \qquad (8.12)$$

where ϕ is defined by Equation (8.8).

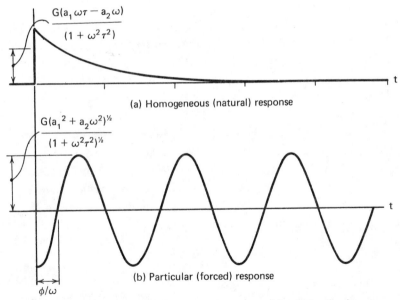

Figure 8.1 Typical natural and forced response of first order circuits

The first term on the right-hand side of Equation (8.12) is the natural response of the system. The second term on the right-hand side is the forced response of the

system. Figure 8.1 is a graphical representation of the
natural and forced responses. The natural response
(Figure 8.1a) is merely a decaying exponential function.
As we have observed previously this function is less than
5% of its initial value after three time constants
($t = 3\tau$). For times greater than this, therefore, the
contribution of the natural response is negligible. The
forced response (Figure 8.1b) has the form of a pure sine
wave. It is, however, displaced in time by ϕ/ω as
indicated in Equation (8.12).

Figure 8.2 shows the complete response and the input
waveform. Initially, the complete response is affected
by the natural response. This initial period is sometimes
referred to as the transient period. After the natural
contribution becomes small both the input and output are
sine waves of the same frequency. This region is known as
the sinusoidal steady state. In most cases when there is
a sine wave input it is a continuing function that is
maintained for a long time. We are, therefore, generally
concerned only with the forced portion of the response.
Indeed, in most cases, the system has settled into the
sinusoidal steady state before measurements or observations are made.

In the sinusoidal steady state the output sine wave
differs from the input sine wave in phase, ϕ, and amplitude. The magnitude, M, expresses the ratio of output
amplitude to input amplitude. The sinusoidal steady
state is then completely described by M and ϕ. For the
first order circuit these parameters have the values:

$$M = \frac{(a_1^2 + a_2^2 \omega^2)^{\frac{1}{2}}}{(1 + \omega^2 \tau^2)^{\frac{1}{2}}} \qquad (8.13a)$$

$$\phi = \tan^{-1} \frac{a_2\omega - a_1\omega\tau}{a_1 + a_2\omega^2\tau} \qquad (8.13b)$$

These magnitude-frequency and phase-frequency relations
are called the frequency response.

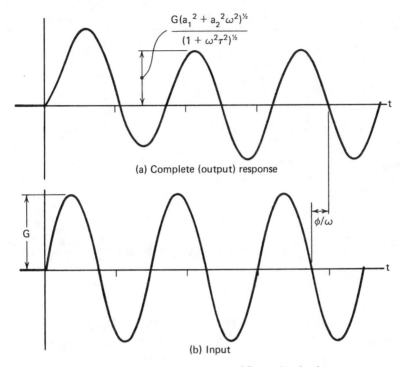

$$\frac{G(a_1{}^2 + a_2{}^2\omega^2)^{1/2}}{(1 + \omega^2\tau^2)^{1/2}}$$

(a) Complete (output) response

ϕ/ω

G

(b) Input

Figure 8.2 Output and input of first order circuit

There is no difference in frequency response between a sine wave input and a cosine wave input. We may think of the cosine input as a sine wave displaced by $t = -\pi/2\omega$. In this case the output will also be displaced by the same time period. The magnitude and phase of output relative to input, however, remain unchanged.

The following examples will demonstrate the frequency response of some first-order circuits.

Example 8.1

Figure 8.3 shows a first-order electrical RC circuit. If the supply voltage has the form $e_s(t) = E \cos\omega t$ find:

(a) the complete voltage response, $e(t)$,
(b) the sinusoidal steady state response,
(c) the frequency response.

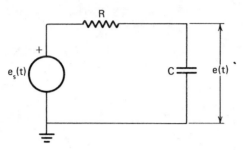

Figure 8.3 First order electrical circuit (example 8.1)

The differential equation that represents the RC series circuit is:

$$RC \frac{de}{dt} + e = e_s(t) = E \cos\omega t$$

$$e(0^+) = 0$$

The complete solution for the output voltage, $e(t)$, consists of the summation of the homogeneous solution and the particular solution. The homogeneous solution for a first-order circuit always has the form:

$$e_H = A_1 e^{-t/RC}$$

To find the particular solution we assume that:

$$e_P = A_2 \sin\omega t + A_3 \cos\omega t$$

Now the substitution of the assumed particular solution into the differential equation yields:

$$[A_2 - \omega RCA_3]\sin\omega t + [\omega RCA_2 + A_3]\cos\omega t = E \cos\omega t$$

If the coefficients of like terms are equated the result is:

$$A_2 - \omega RC A_3 = 0$$

$$\omega RC\ A_2 + A_3 = E$$

The solution of these simultaneous equations provides A_2 and A_3 for the particular solution. The result is:

$$e_p = \frac{E}{[1 + (\omega RC)^2]} [\omega RC\ \sin\omega t + \cos\omega t]$$

$$= \frac{E}{[1 + (\omega RC)^2]^{\frac{1}{2}}} \cos(\omega t - \phi)$$

where

$$\phi = + \tan^{-1} \omega RC \qquad \cos\phi = \frac{1}{\sqrt{1 + (\omega RC^2)}}$$

The complete solution for the voltage across the capacitor then has the form:

$$e(t) = A_1\ e^{-t/RC} + \frac{E}{[1 + (\omega RC)^2]^{\frac{1}{2}}} \cos(\omega t - \phi)$$

The constant A_1 may now be determined from the initial condition with the result that the complete solution (part a) is:

$$e(t) = \frac{- E\ e^{-t/RC}\ \cos\phi + E\ \cos(\omega t - \phi)}{\sqrt{1 + (\omega RC)^2}}$$

$$e(t) = \frac{- E\ e^{-t/RC}}{[1 + (\omega RC)^2]} + \frac{E\ \cos(\omega t - \phi)}{[1 + (\omega RC)^2]^{\frac{1}{2}}}$$

To obtain the sinusoidal steady state response we should recognize that the exponential term in the complete response approaches zero. Therefore, the sinusoidal steady state response is the same as the forced (particular) response and the answer to part b is:

$$e(t) = \frac{E}{[1 + (\omega RC)^2]^{\frac{1}{2}}} \cos(\omega t - \phi)$$

The frequency response (part c) is:

$$M = \frac{E}{[1 + (\omega RC)^2]^{\frac{1}{2}}} \; (\frac{1}{E}) = \frac{1}{[1 + (\omega RC)^2]^{\frac{1}{2}}}$$

This same result could have been reached by using the general expression for the frequency response of a first order system (Equation (8.13)) with $a_1 = 1$, $a_2 = 0$, and $\tau = RC$. The parameter ωRC represents a normalized angular frequency. Figure 8.4 shows the magnitude, M and phase, ϕ, plotted against this normalized frequency. The scale on the left represents magnitude. At zero frequency the capacitor has infinite impedance and the voltage of the output is equal to the input voltage. The magnitude is, therefore, equal to unity for dc signals. As the frequency increases the output continually decreases. At very high frequencies the magnitude approaches zero. Here the capacitor behaves as a low impedance and the output is near ground potential. The scale on the right represents the phase shift in degrees. At low frequencies the input and output are almost in phase. At the frequency $\omega = 1/RC$ the phase lags by 45 degrees. This series RC circuit is often referred to as a first order lag because of its phase-frequency characteristics.

As mentioned previously the frequency response is the same for sine or cosine inputs. However, the transient portion of the complete response (the constant A_1) is not equal for both input waveforms.

Second-Order Circuits

The most general differential equation for linear time invariant second-order circuits is given in Equation (6.1) as:

$$\frac{d^2 y}{dt^2} + 2\delta\omega_N \frac{dy}{dt} + \omega_N^2 y = a_1 g(t) + a_2 \frac{dg(t)}{dt} + a_3 \frac{d^2 g(t)}{dt^2} \quad (8.14)$$

When the input function is a sine wave $g(t) = G \sin\omega t$ and Equation (8.14) becomes:

$$\frac{d^2y}{dt^2} + 2\delta\omega_N \frac{dy}{dt} + \omega_N^2 y = Gd_1 \sin\omega t + Gd_2 \cos\omega t \quad (8.15)$$

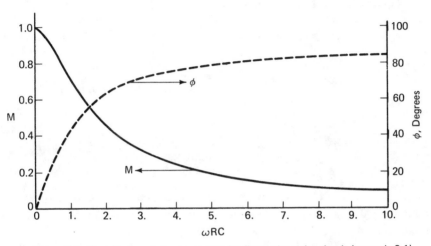

Figure 8.4 Magnitude and phase response for first order series circuit (example 8.1)

where $d_1 = (a_1 - a_3\omega^2)$ and $d_2 = a_2\omega$. The complete solution for Equation (8.15) is the sum of the homogeneous and particular solutions. In Chapter VI, we learned that there are three possible homogeneous solutions for the general second-order circuit. The solution that is appropriate depends on whether the circuit is underdamped, overdamped, or critically damped. For an underdamped or overdamped circuit the homogeneous solution has the form given in Equation (6.6) or (6.8):

$$Y_H = A_1 e^{r_1 t} + A_2 e^{r_2 t} \quad (8.16)$$

where r_1 and r_2 are the roots of the characteristic equation. To determine the particular solution we once again apply the method of undetermined coefficients. We, therefore, assume as in Equation (8.3) a particular solution, y_p, that is:

$$y_p = A_3 \sin\omega t + A_4 \cos\omega t \qquad (8.17)$$

If Equation (8.17) is substituted into Equation (8.15) the result is:

$$[c_1 A_3 - c_2 A_4]\sin\omega t + [c_2 A_3 + c_1 A_4]\cos\omega t$$
$$= Gd_1 \sin\omega t + Gd_2 \cos\omega t \qquad (8.18)$$

where

$$c_1 = (\omega_N^2 - \omega^2) \quad \text{and} \quad c_2 = 2\delta\omega\omega_N$$

Now if the coefficients of like terms on each side of Equation (8.18) are equated we obtain the simultaneous equations:

$$c_1 A_3 - c_2 A_4 = Gd_1$$
$$c_2 A_3 + c_1 A_4 = Gd_2 \qquad (8.19)$$

The coefficients A_3 and A_4 are determined from Equation (8.19). The particular solution (Equation (8.17)) thus becomes:

$$y_p = \frac{G}{c_1^2 + c_2^2}[(d_1 c_1 + d_2 c_2)\sin\omega t + (d_2 c_1 - d_1 c_2)\cos\omega t]$$
$$= \frac{G(d_1^2 + d_2^2)^{\frac{1}{2}}}{(c_1^2 + c_2^2)^{\frac{1}{2}}} \sin(\omega t + \phi) \qquad (8.20)$$

where

$$\phi = \tan^{-1}\frac{d_2 c_1 - d_1 c_2}{d_1 c_1 + d_2 c_2}$$

The complete solution is the sum of Equations (8.16) and (8.20).

$$y = A_1 \, e^{r_1 t} + A_2 \, e^{r_2 t} + \frac{G(d_1^2 + d_2^2)^{\frac{1}{2}}}{(c_1^2 + c_2^2)^{\frac{1}{2}}} \sin(\omega t + \phi) \quad (8.21)$$

The constants A_1 and A_2 are now determined from the initial conditions. However, when we are concerned only with the sinusoidal steady state we need not find A_1 and A_2. In the sinusoidal steady state (t approaching infinity), the effects of the exponential terms in Equation (8.21) fade away. As a result, the sinusoidal steady state and the particular solution (Equation (8.20)) are the same. Thus, the frequency response relations for a second order circuit are:

$$M = \left[\frac{d_1^2 + d_2^2}{c_1^2 + c_2^2} \right]^{\frac{1}{2}} = \left[\frac{(a_1 - a_3 \omega^2) + (a_2 \omega)^2}{(\omega_N^2 - \omega^2)^2 + (2\delta \omega \omega_N)^2} \right]^{\frac{1}{2}} \quad (8.22)$$

$$\phi = \tan^{-1} \frac{d_2 c_1 - d_1 c_2}{d_1 c_1 + d_2 c_2} = \tan^{-1} \frac{(a_2 \omega)(\omega_N^2 - \omega^2) - (a_1 - a_3 \omega^2)(2\delta \omega \omega_N)}{(a_1 - a_3 \omega^2)(\omega_N^2 - \omega^2) + (a_2 \omega)(2\delta \omega \omega_N)}$$

Equation (8.22) expresses the frequency response for all second order circuits. The magnitude response given in Equation (8.22) gets large without bound when the frequency of the input sine wave is the same as the natural frequency of the circuit ($\omega = \omega_N$) and the circuit has no damping ($\delta = 0$). Both conditions are required to produce this result. From a mathematical viewpoint the particular and homogeneous solutions under these conditions contain identical terms. Thus to derive the complete solution we must multiply the assumed particular solution by time. From a physical viewpoint these conditions are equivalent to exciting an undamped circuit with its own natural frequency. In theory the output amplitude of oscillation would grow without bound. As a practical matter, however,

the component models would no longer be linear and the
amplitude of the oscillations would settle at some large
value (limit cycle) provided that the system is not
damaged in the process.

The following example demonstrates the frequency response
of a specific second order circuit. Other second order
circuits may be treated in the same way.

Example 8.2

A pneumatic signal generator supplies air to a closed
chamber through a long line (Figure 8.5a). The pressure
waveform of the generator is $p(t) = P \cos\omega t$. If the
chamber has a capacitance, C, and the line has resistance,
R, and inertance, L, determine the frequency response of
the flow, q.

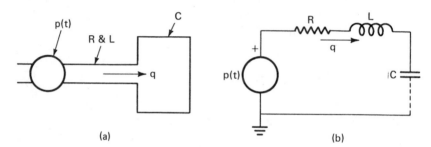

(a)   (b)

Figure 8.5   Fluid circuit (example 8.2)

Figure 8.5b shows the analogous circuit diagram for
the fluid circuit. The differential equation representing the flow is:

$$\frac{d^2q}{dt^2} + \frac{R}{L}\frac{dq}{dt} + \frac{1}{LC}q = \frac{1}{L}\frac{dp(t)}{dt} = -\frac{\omega P}{L}\sin\omega t$$

Since we want to find the frequency response we need
only consider the particular solution, q_p. If we
assume that:

$$q_p = A_1 \sin\omega t + A_2 \cos\omega t$$

then the substitution into the differential equation yields:

$$[(\frac{1}{LC} - \omega^2)A_1 - \frac{\omega R}{L}A_2]\sin\omega t$$

$$+ [\frac{\omega R}{L}A_1 + (\frac{1}{LC} - \omega^2)A_2]\cos\omega t = -\frac{\omega P}{L}\sin\omega t$$

If we equate the coefficients of the sine and cosine terms on each side of the equation we obtain that:

$$q_P = \frac{\omega P/L}{[(\frac{1}{LC}-\omega^2)^2+(\frac{\omega R}{L})^2]}[-(\frac{1}{LC}-\omega^2)\sin\omega t+(\frac{\omega R}{L})\cos\omega t]$$

$$= \frac{\omega P/L}{[(\frac{1}{LC}-\omega^2)^2+(\frac{\omega R}{L})^2]^{\frac{1}{2}}}\cos(\omega t + \phi)$$

where

$$\phi = \tan^{-1}\frac{(\frac{1}{LC} - \omega^2)}{\omega R/L}$$

Now by definition:

$$\omega_N = (\frac{1}{LC})^{\frac{1}{2}} \quad \text{and} \quad \delta = \frac{R}{2L\omega_N}$$

so that:

$$q_P = \frac{\omega P/L}{[(\omega_N^2 - \omega^2)^2 + (2\delta\omega\omega_N)^2]^{\frac{1}{2}}}\cos(\omega t + \phi)$$

Thus:

$$M = \frac{\omega/L}{[(\omega_N^2-\omega^2)^2+(2\delta\omega\omega_N)^2]^{\frac{1}{2}}} = \frac{(C/L)^{\frac{1}{2}}(\omega/\omega_N)}{[1+(4\delta^2-2)(\omega/\omega_N)^2+(\omega/\omega_N)^4]^{\frac{1}{2}}}$$

$$\phi = \tan^{-1}\frac{[1 - (\omega/\omega_N)^2]}{2\delta(\omega/\omega_N)}$$

In this case the magnitude function is not dimension-less since the input is a pressure and the output is a

flow.   However, the parameter $(L/C)^{\frac{1}{2}}$ M is dimensionless.
Figure 8.6 shows this parameter plotted against the
normalized frequency ω/ω_N for various values of the
damping factor, δ.   This is the magnitude-frequency
response for this circuit.   Note that the response has
a peak at the natural frequency $(\omega/\omega_N = 1)$ when the
damping factor is small.

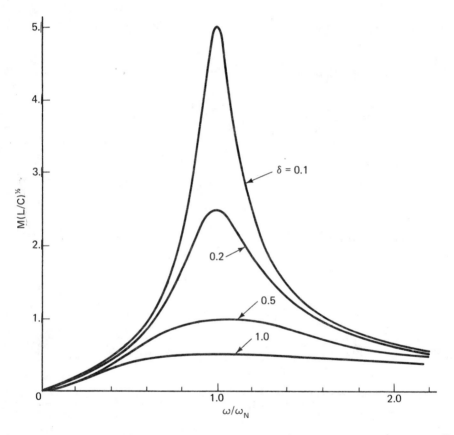

Figure 8.6 Magnitude-frequency response (example 8.2)

Figure 8.7 shows the phase-frequency response.   As the
frequency increases the phase goes from a 90 degree lead
to a 90 degree lag.   When the input frequency is the
same as the natural frequency $(\omega/\omega_N = 1)$ the input and

output are in phase. Although we present magnitude and
phase (Figures 8.6 and 8.7) to indicate frequency
response one relation or the other is usually sufficient.
In the circuits we deal with in this text (minimum
phase circuits) the shape of one plot is evidence of
the shape of the other. Thus, for example, when phase
changes rapidly in a narrow frequency band the accom-
panying amplitude response has a large peak. Conversely
a large peak on the magnitude plot means a rapid phase
change on the phase plot.

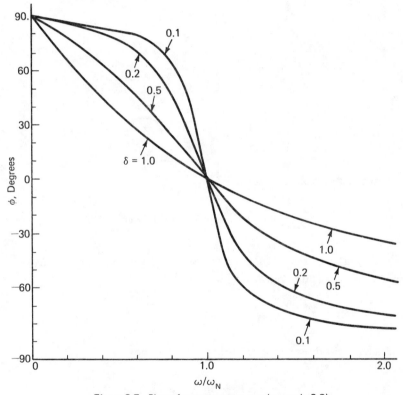

Figure 8.7 Phase-frequency response (example 8.2)

The results presented in this example could have been
determined directly from Equation (8.22) with $a_2 = 1/L$
and $a_1 = a_3 = 0$.

8.2 SINUSOIDAL STEADY STATE

The procedure just described to obtain the total response
to a sinusoidal input is very cumbersome as far as the
forced or steady-state response is concerned.   If we recall
from Chapter VII that a sinusoid may be represented by the
real or imaginary part of $e^{j\omega t}$, the solution is obtained
much more rapidly.

Example 8.3

Figure 8.8a shows a mechanical mass-damper circuit.
If the applied force is $f(t) = F \cos\omega t$, find the steady-
state velocity of the mass.

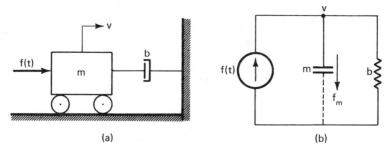

(a)   (b)

Figure 8.8 Mechanical circuit (example 8.3)

The analogous circuit is shown in Figure 8.8b.   The
equilibrium equation is:

$$m \frac{dv}{dt} + bv = \text{Re } [F \ e^{j\omega t}]$$

and the steady-state velocity must have the form:

$$v(t) = \text{Re } [V \ e^{j\omega t}]$$

Using the complete driving function $F \ e^{j\omega t}$ and substituting gives:

$$j\omega mV + bV = F$$

from which:

$$V = \frac{F}{b + j\omega m} = \frac{F}{Y e^{j\phi}}$$

where

$$Y = \sqrt{b^2 + (\omega m)^2} \quad \text{and} \quad \tan\phi = \frac{\omega m}{b}$$

That is:

$$v(t) = \text{Re}\left[\frac{F}{Y e^{j\phi}} e^{j\omega t}\right]$$

$$= \frac{F}{Y} \text{Re}[e^{j(\omega t-\phi)}] = \frac{F}{Y} \cos(\omega t-\phi)$$

Note that if only the amplitude ($M = F/Y$) and phase (ϕ) of the steady-state response are required, the final part of this solution is quite unnecessary. Figure 8.9 shows a plot of the magnitude and phase response for this mechanical circuit.

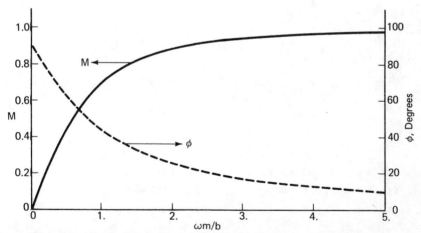

Figure 8.9 Magnitude and phase response for mechanical circuit (example 8.3)

Although the procedure used in this example is more convenient than that in Section 8.1, it is possible to improve it still further by noting the nature of a system

function. Suppose, for example, that the input to a
circuit has the form $x(t) = X e^{st}$ and the output has the
form $y(t) = Y e^{st}$. The system function is, by definition:

$$G(s) = \frac{Y}{X} \qquad (8.23)$$

For the sinusoidal steady-state, we have already noted
that we are dealing with the particular case of an exponential function where $s = j\omega$. This is an application of
Euler's Equation (Equation (2.28)):

$$e^{j\omega t} = \cos\omega t + j \sin\omega t$$

Thus when input and output are cosine waves we may write:

$$x(t) = Re[X e^{j\omega t}] \qquad (8.24a)$$

$$y(t) = Re[Y e^{j\omega t}] \qquad (8.24b)$$

If Equation (8.23) is substituted into Equation (8.24b)
the result is:

$$y(t) = Re[X G(j\omega)e^{j\omega t}] \qquad (8.25)$$

   In terms of cosine functions Equation (8.25) is equivalent to:

$$y(t) = X|G(j\omega)| \cos(\omega t + \phi) \qquad (8.26)$$

where $|G(j\omega)|$ is the magnitude of the transfer function
and ϕ is the angle of $G(j\omega)$. The amplitude ratio of output, $y(t)$ to input $x(t)$ is therefore:

$$M = \frac{X|G(j\omega)|}{X} = |G(j\omega)| \qquad (8.27)$$

and the phase may be expressed as:

$$\phi = \underline{/G(j\omega)} \qquad (8.28)$$

To find the frequency response magnitude and phase we merely substitute $j\omega$ for s in the transfer function and then perform the operations required in Equations (8.27) and (8.28).

Example 8.4

As an illustration let us reconsider the fluid circuit of Example 8.2. The transfer function $G(s)$ is:

$$G(s) = \frac{Q}{P} = \frac{Cs}{LCs^2 + RCs + 1}$$

If we replace s by $j\omega$ the result is:

$$G(j\omega) = \frac{j\omega C}{(1 - \omega^2 LC) + j\omega RC}$$

From the rules of complex algebra we may express $G(j\omega)$ as:

$$G(j\omega) = \frac{\omega C}{[(1-\omega^2 LC)^2+(\omega RC)^2]^{\frac{1}{2}}} \underline{/90° - \tan^{-1} \frac{\omega RC}{1-\omega^2 LC}}$$

and by algebraic manipulation with $\omega_N = (1/LC)^{\frac{1}{2}}$ and $2\delta\omega_N = R/L$ the above equation becomes:

$$G(j\omega) = \frac{(C/L)^{\frac{1}{2}}(\omega/\omega_N)}{[1+(4\delta^2-2)(\omega/\omega_N)^2+(\omega/\omega_N)^4]^{\frac{1}{2}}} \underline{/\tan^{-1} \frac{[1-(\omega/\omega_N)^2]}{2\delta(\omega/\omega_N)}}$$

and we can recognize the amplitude and phase of $G(j\omega)$ as M and ϕ from Example 8.2.

Example 8.5

The mass-spring damper system shown in Figure 8.10a is subjected to an oscillatory force $f(t) = F \cos\omega t$, N. Find the amplitude and phase of the forces in each element.

The analogous circuit is shown in Figure 8.10b from which the equilibrium equation is obtained:

$$(ms + b + \frac{k}{s})V = F$$

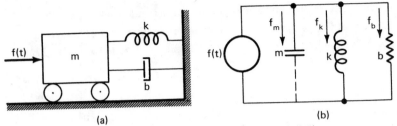

Figure 8.10 Mechanical circuit (example 8.5)

The solution is:

$$V = \frac{Fs}{ms^2 + bs + k}$$

and the required forces are:

$$F_m = msV = \frac{mFs^2}{ms^2+bs+k} \quad ; \quad F_b = bV = \frac{bFs}{ms^2+bs+k}$$

$$F_k = \frac{k}{s} V = \frac{kF}{ms^2+bs+k}$$

Replacing s by $j\omega$ gives:

$$F_m(j\omega) = \frac{-\omega^2 mF}{(k-\omega^2 m)+j\omega b} \quad ; \quad F_b(j\omega) = \frac{j\omega bF}{(k-\omega^2 m)+j\omega b}$$

$$F_k(j\omega) = \frac{kF}{(k-\omega^2 m)+j\omega b}$$

Thus:

$$F_m(j\omega) = \frac{\omega^2 mF}{\sqrt{(k-\omega^2 m)^2+(\omega b)^2}} \; \underline{/180°-\phi}$$

$$F_b(j\omega) = \frac{\omega bF}{\sqrt{(k-\omega^2 m)^2+(\omega b)^2}} \; \underline{/90°-\phi}$$

$$F_k(j\omega) = \frac{kF}{\sqrt{(k-\omega^2 m)^2+(\omega b)^2}} \; \underline{/-\phi}$$

where:

$$\tan\phi = \frac{\omega b}{k - \omega^2 m}$$

Normally this form is adequate and with experience does convey the complete information on the responses. However, if the time functions are required it is simply a matter of noting that since the driving functions was $F \cos\omega t$, the responses must be:

$$f_m(t) = \frac{\omega^2 mF}{\sqrt{(k-\omega^2 m)^2 + (\omega b)^2}} \cos(\omega t + 180 - \phi)$$

The magnitude and phase responses for this mechanical circuit are shown in Figure 8.11 for the case where $b = \sqrt{km}$.

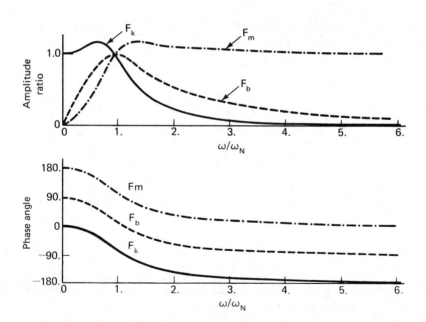

Figure 8.11 Magnitude and phase frequency response (example 8.5)
$(b = \sqrt{km})$

8.3 POWER IN THE SINUSOIDAL STEADY-STATE

For all the physical systems we have considered except
thermal systems, the through and across variables were
chosen so that not only would the equilibrium equations
have the same mathematical form but also their product
would be power. In the case of the dissipative elements,
this power normally appears as heat. In the case of the
storage elements the power is the rate at which they are
storing energy. If we use the polarity convention con-
sistently, the power flow into any of the passive elements
is given by:

$$P(t) = [AV(t)][TV(t)] \qquad (8.29)$$

where $AV(t)$ and $TV(t)$ are the instantaneous values of
the across and through variables respectively.

In the sinusoidal steady-state both AV and TV are
sinusoids and thus the power flow is not constant. For
example, the power dissipated in a resistance is:

$$P(t) = EI \cos^2 \omega t \qquad (8.30)$$

if

$$e(t) = E \cos \omega t \quad \text{and} \quad i(t) = I \cos \omega t$$

This expression must be modified in order to interpret it
and it becomes, by use of the double angle relation:

$$P(t) = \frac{EI}{2}(1 + \cos 2\omega t) \qquad (8.31)$$

Thus, the power flow is oscillating at twice the frequency
of the current and voltage, but there is a time average
value given by:

$$P_{av} = \frac{EI}{2} \qquad (8.32)$$

The factor of 1/2 has usually been considered troublesome
and may be eliminated by working with effective values of
current and voltage so that:

$$P_{av} = E_{eff} I_{eff} \qquad (8.33)$$

where:

$$E_{eff} = \frac{E}{\sqrt{2}} \quad \text{and} \quad I_{eff} = \frac{I}{\sqrt{2}}$$

The effective values are also known as the rms values
since more generally they can be shown to be the square
root of the average of the square of the current or voltage. In much of the literature Equation (8.33) does not
have any subscripts and appears simply as $P = EI$. Care
is therefore required when writing and interpreting
expressions involving power in the sinusoidal steady
state. In addition it may be noted that $Z(j\omega) = E/I =
E_{eff}/I_{eff}$. Therefore, all equations involving impedances
or admittances may be used with either effective values
or amplitudes.

When one or more storage elements is present, the
expression for power must be modified still further.
If we consider a simple series circuit consisting of a
resistor and an inductor, we have:

$$Z(j\omega) = R + j\omega L$$

and hence

$$I(j\omega) = \frac{E}{R+j\omega L} = \frac{E}{|Z|} \underline{/-\phi}$$

where

$$|Z| = \sqrt{R^2 + (\omega L)^2} \quad \text{and} \quad \tan\phi = \frac{\omega L}{R}$$

Thus, if $e(t) = E \cos\omega t$, the current must be given by:

$$i(t) = \frac{E}{|Z|} \cos(\omega t - \phi)$$

and the power is:

$$P(t) = EI \cos\omega t \cos(\omega t - \phi)$$

$$= \frac{EI}{2} [\cos\phi + \cos(2\omega t - \phi)] \qquad (8.34)$$

Again the time variation in power is at twice the excitation frequency but the average value is now:

$$P_{av} = \frac{EI}{2} \cos\phi = E_{eff} I_{eff} \cos\phi \qquad (8.35)$$

The ratio of the average power to the product of current and voltage is usually called the power factor which in this case is the cosine of the phase angle ϕ.

Example 8.6

If we return to Example 8.3 where F is the amplitude of the applied force, determine the power dissipated. The velocity amplitude and phase were found to be:

$$V = \frac{F}{\sqrt{b^2 + (\omega m)^2}} \quad \text{and} \quad \phi = \tan^{-1} \frac{\omega m}{b}$$

Thus, the average power entering the system is:

$$P = V_{eff} F_{eff} \cos\phi$$

$$= \left[\frac{F^2_{eff}}{\sqrt{b^2 + (\omega m)^2}} \right] \left[\frac{b}{\sqrt{b^2 + (\omega m)^2}} \right]$$

$$= \frac{F^2_{eff} b}{b^2 + (\omega m)^2} = V^2_{eff} b$$

Since an ideal mass may only store energy, the average power associated with it must be zero and hence it is essential that any expression for the average power entering a system should equal the average power dissipated in the dissipative elements. Also, it can be noted that the expressions I^2R and E^2/R and their analogs may be used as correct power relations provided I is the effective value of the current flowing through the resistance R or E is the effective value of the voltage across it.

Maximum Power Transfer

There are occasions when it is necessary to choose a dissipative element so that the power dissipated in it is the maximum possible with a given source. In Figure 8.12, $E_{e(eff)}$ and Z_e may be considered as the Thevenin equivalent of any circuit. For the sinusoidal steady state, $Z_e(s)$ becomes $Z_e(j\omega)$ and at any one frequency this can be simplified into a real part (equivalent resistance) and an imaginary part (equivalent reactance), $R_e + j\, X_e$. Note that the values of R_e and X_e usually depend on the frequency.

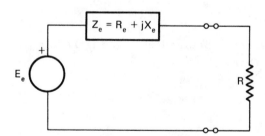

Figure 8.12 Circuit reduction for maximum power transfer

Hence the current can be expressed, using effective values as:

$$I_{eff}(j\omega) = \frac{E_{e(eff)}}{R_e + jX_e + R}$$

The power dissipated in R is:

$$P = \frac{E_e^2(eff)R}{(R+R_e)^2 + X_e^2} \qquad (8.36)$$

The maximum value of P is found in the usual manner by
differentiating with respect to R, equating to zero and
solving for R.   The result is that power is maximum when:

$$R = \sqrt{R_e^2 + X_e^2} \qquad (8.37)$$

That is maximum power is dissipated in R when it is numer-
ically the same as the magnitude of $Z_e(j\omega)$.   For the
particular case where $X_e = 0$ the condition of $R = R_e$ gives
maximum power transfer.

If, in addition to adjusting the value of R, it is
possible to have either a capacitor or an inductor connected in series with the resistor, the condition for
maximum power is modified.   That is, the current becomes:

$$I_{eff}(j\omega) = \frac{E_e(eff)}{Z_e(j\omega) + Z(j\omega)}$$

$$= \frac{E_e(eff)}{(R_e+R) + j(X_e+X)} \qquad (8.38)$$

and maximum power is obtained when Z is the complex conjugate of Z_e.   That is:

$$R = R_e \ , \quad X = - X_e \qquad (8.39)$$

Example 8.7

A mass-damper system is shown in Figure 8.13a.   It is
subjected to an oscillatory force of frequency 1.592 Hz
and amplitude 12.5 N.   If $m_1 = 3$ kg, $m_2 = 0.96$ kg and
$b_1 = 40$ N·s/m, determine (a) the damping required to
dissipate maximum power in b_2 and (b) the maximum power
dissipated.

The analogous circuit is shown in Figure 8.13b and the
Norton equivalent of the "fixed" part of the circuit is
obtained from Figure 8.13c, the values being:

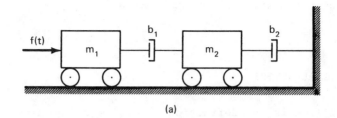

(a)

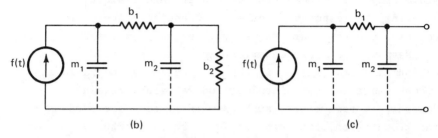

(b)   (c)

Figure 8.13 Mechanical circuit (example 8.7)

$$F_e = \frac{b_1 F}{b_1 + m_1 s} = \frac{40F}{40 + j30}$$

$$Y_e = m_2 s + \frac{1}{1/m_1 s + 1/b_1} = m_2 s + \frac{bm_1 s}{m_1 s + b_1}$$

$$= j9.6 + \frac{j(40)(30)}{40+j30} = 14.4 + j28.8$$

For maximum power dissipation in the second damper, it must be set to the value $b_2 = \sqrt{14.4^2+28.8^2}=32.2$ N·s/m. The total admittance is the sum of b_2 and Y_e. Thus:

$$Y = 46.6 + j28.8 = 54.8 \underline{/31.7°}$$

If the amplitude of the driving force is 12.5 N:

$$F_e = \frac{500}{40+j30} = 8 - j6 = 10\underline{/-36.9°}$$

The amplitude of the velocity across the damper is:

$$V = \frac{|F_e|}{|Y|} = \frac{10}{54.8} = 0.182 \text{ m/s}$$

Thus the maximum power dissipated is:

$$P = V_{eff}^2 \, b_2 = \frac{V^2}{2} \, b_2 = 0.533 \text{ W}$$

8.4 SIGNAL FILTERS

A filter is, by definition, a device for maximizing or
minimizing the response to certain frequencies without
altering the response to other frequencies. The extent
to which a particular frequency is present or absent in
the response is measured primarily by its amplitude. The
performance of a filter is then indicated by the ratio of
output amplitude to input amplitude at each frequency.
However, in some cases, such as when filters are used in
instrumentation, the phase shift is equally important.

Although the theory and application of electrical signal
filters is the most well-known, the same principles are
relevant for the other physical media. Thus there are
mechanical circuits to filter mechanical signals and fluid
circuits to filter fluid signals.

We consider only passive filters here. These are
circuits that are constructed with the passive components
(resistance, capacitance and inductance).

Low-Pass Filters

A low-pass filter, as the name implies, passes low frequency signals and blocks high frequency signals. Figure
8.14 shows a schematic representation of the characteristics of low-pass filters. For the ideal low-pass filter
there is a sharp cut-off between the pass region and the
block region. This means that an input signal in the low
frequency pass region produces an output signal of the
same amplitude and, of course, the same frequency. In
other words, the signal passes through the filter. On
the other hand, an input signal in the high frequency
block region yields no output signal. Actual low-pass
filters do not have as sharp a cut-off as ideal low-pass

filters. Some frequencies that are desired eliminated
are passed with greatly reduced amplitudes. Other frequencies that are to be retained are also passed but with
only slightly reduced amplitudes.

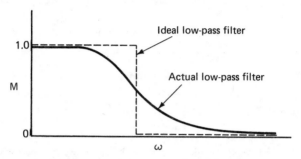

Figure 8.14 Low-pass filter characteristics

The transfer function for typical low-pass passive
filters may be described by:

$$G(s) = \frac{K}{D(s)} \qquad (8.40)$$

where K is a constant less than unity and D(s) is a function of s. The distinguishing feature of the low-pass
filter is that the numerator of the transfer function is
not a function of s. The following examples will illustrate some passive low-pass filter circuits.

Example 8.8

In the mass-damper circuit (Figure 8.15a) the input
velocity is $V_i \sin\omega t$. This circuit may be considered
as a mechanical filter where the output is the velocity
of the mass. In another mass-damper circuit (Figure
8.15b) there are two mass-damper combinations in tandem.
This circuit has the same input velocity as the single
mass-damper circuit. The output, however, is the
velocity of the second mass component. Determine the
magnitude-frequency response for both circuits and
discuss their behavior as mechanical filters.

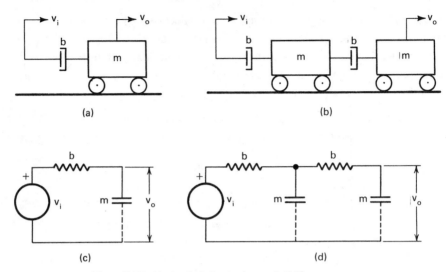

Figure 8.15   Mechanical circuits (example 8.8)

The analogous circuit for the single mass-damper combination is shown in Figure 8.15c. The transfer function $G_1(s)$ is:

$$G_1(s) = \frac{V_o}{V_i} = \frac{1}{1 + ms/b}$$

The magnitude-frequency response from Equation (8.27)
is:

$$M_1 = |G_1(j\omega)| = \left|\frac{1}{1 + j\omega m/b}\right| = \frac{1}{[1 + (\omega m/b)^2]^{\frac{1}{2}}}$$

The analogous circuit for the two-section mass-damper
arrangement is shown in Figure 8.15d. The transfer function, $G_2(s)$ is:

$$G_2(s) = \frac{V_o}{V_i} = \frac{1}{1 + 3ms/b + (ms/b)^2}$$

The magnitude-frequency response, is therefore:

$$M_2 = |G_2(j\omega)| = \left|\frac{1}{[1-(\omega m/b)^2]+j3\omega m/b}\right|$$

$$= \frac{1}{([1-(\omega m/b)^2]^2+9(\omega m/b)^2)^{\frac{1}{2}}}$$

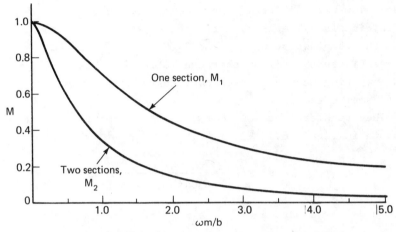

Figure 8.16 Magnitude-frequency curves (example 8.8)

Figure 8.16 shows the magnitude-frequency curves for M_1 and M_2. Both circuits may be classified as low-pass filters. The one section circuit passes larger amplitude output signals at all frequencies than the two-section circuit. This is a desirable feature in the low frequency region but is undesirable at high frequencies. For example, suppose the desired cut-off frequency is $\omega = b/m$. At frequencies above cut-off ($\omega m/b > 1$) the two-section circuit provides considerably more attenuation than the one-section circuit. In fact, the two-section arrangement practically blocks out all frequencies above $\omega m/b = 3$. However, below cut-off ($\omega m/b < 1$) the two-section circuit still attenuates signals although no attenuation is desired.

When additional mass-damper sections are employed the magnitude-frequency curves have sharper cut-offs but the cut-off frequency decreases.

Example 8.9

The electric circuit shown in Figure 8.17 may be used as a low-pass filter if the RLC components are selected judiciously. The characteristics of the filter depend on the damping factor $\delta = \sqrt{L}/(2R\sqrt{C})$ and $\omega_N = 1/\sqrt{LC}$. Investigate the filter characteristics of this circuit by plotting magnitude-frequency curves for various values of damping factor.

The transfer function, $G(s)$, of the electric circuit shown in Figure 8.17 is:

$$G(s) = \frac{E_O}{E_i} = \frac{1}{LCs^2 + \frac{L}{R} s + 1}$$

from which we may develop the magnitude as:

$$M = |G(j\omega)| = \left| \frac{1}{(1-\omega^2 LC)+j\omega L/R} \right|$$

Now, in terms of the normalizing natural frequency, ω_N and damping factor, δ

$$M = \left| \frac{1}{(1-\omega^2/\omega_N^2)+j\ 2\delta(\omega/\omega_N)} \right|$$

$$= \frac{1}{([1-(\omega/\omega_N)^2]^2+4\delta^2(\omega/\omega_N)^2)^{\frac{1}{2}}}$$

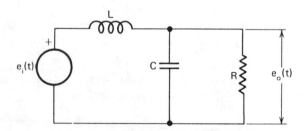

Figure 8.17   Electrical circuit (example 8.9)

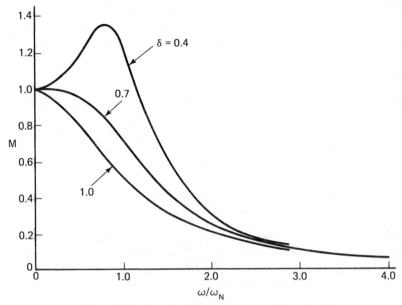

Figure 8.18 Magnitude-frequency curves (example 8.9)

Figure 8.18 shows a plot of magnitude versus non-dimensional frequency ω/ω_N for various values of δ. The circuit attenuates high frequency (i.e. $\omega/\omega_N > 2$) and may be classed as a low-pass filter. However, when the damping factor is small the circuit may magnify some signals at lower frequencies. A compromise must be made between sharp cut-off with low frequency magnification and gradual cut-off without magnification. Damping factors of approximately 0.6 often provide a good trade-off.

High-Pass Filters

A high-pass filter has characteristics opposite to a low-pass filter. It passes high frequency signals and blocks low frequency signals. Figure 8.19 shows the characteristic of an ideal high-pass filter and a typical actual high-pass filter. The ideal filter has a sharp discontinuity in its characteristic. For frequencies higher than the discontinuity the magnitude of output

and input are equal. The output has no magnitude when
the frequencies are lower than the cut-off frequency. In
the actual filter, the very low frequency and very high
frequency behavior is the same as the ideal filter. However, at intermediate frequencies there is a gradual
magnitude change rather than the abrupt change desired.

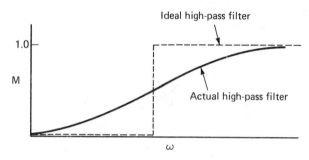

Figure 8.19 High-pass filter characteristics

The transfer function for typical high-pass passive
filters has the form:

$$G(s) = \frac{A_1 s^m}{A_1 s^m + A_2 s^{m-1} + \ldots A_m s + A_{m+1}} \qquad (8.41)$$

where A_1, A_2 ... are the coefficients of the polynomial
in s. The distinguishing feature of the high-pass transfer
function is that the numerator term has the same order as
the highest order term in the denominator. Examples 8.10
and 8.11 which follow will describe some high-pass filter
circuits.

Example 8.10

Figures 8.20a and b show one-section and two-section
spring-damper combinations. The values of the damping,
b, and the spring constant, k, are the same in each
mechanical circuit. The velocity input to the damper
is V_i sinωt in both circuits. The location of the out-

put velocity is indicated in Figures 8.20a and b. Determine the magnitude-frequency characteristics of these
circuits and discuss the characteristics in terms of
signal filters.

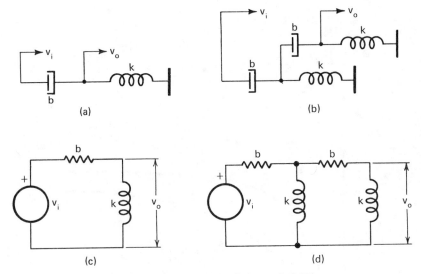

Figure 8.20 Mechanical circuit (example 8.10)

The analogous circuit diagrams for the one-section and
two-section mechanical circuits are shown in Figures
8.20c and d. The transfer function $G_1(s)$, of the single
spring-damer circuit is:

$$G_1(s) = \frac{bs/k}{bs/k + 1}$$

The magnitude-frequency response is:

$$M_1 = |G_1(j\omega)| = \left|\frac{j(\omega b/k)}{j(\omega b/k) + 1}\right|$$

$$= \frac{(\omega b/k)}{[1 + (\omega b/k)^2]^{\frac{1}{2}}}$$

For the two section spring-damper circuit the transfer function, $G_2(s)$, is:

$$G_2(s) = \frac{b^2 s^2 / k^2}{b^2 s^2 / k^2 + 3bs/k + 1}$$

and the frequency response is:

$$M_2 = |G_2(j\omega)| = \left| \frac{- (\omega b/k)^2}{(1 - [\omega b/k]^2) + j3\omega b/k} \right|$$

$$= \frac{(\omega b/k)^2}{[(1 - [\omega b/k]^2)^2 + 9(\omega b/k)^2]^{\frac{1}{2}}}$$

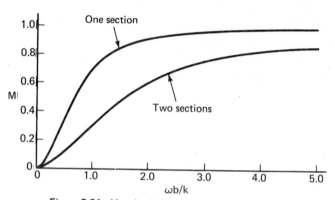

Figure 8.21 Magnitude-frequency curves (example 8.10)

Figure 8.21 shows the magnitude-frequency plots for M_1 and M_2. The non-dimensional frequency parameter for this circuit is $\omega b/k$. The results indicate that both circuits pass very high frequency signals and block very low frequency signals. Thus we may classify these circuits as types of high-pass filter. The two-section circuit does a better job at eliminating the low frequencies. However, it attenuates the high frequencies in the process. Once again, we have to make a trade-off to obtain the most satisfactory circuit.

Example 8.11

Investigate the magnitude-frequency characteristics of the LC electrical circuit shown in Figure 8.22. The input is a sine wave and the output is the voltage across the inductance.

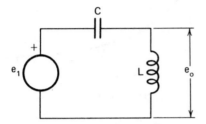

Figure 8.22 Electric circuit (example 8.11)

The transfer function of the LC circuit is:

$$G(s) = \frac{LCs^2}{LCs^2 + 1}$$

The magnitude is therefore:

$$M = |G(j\omega)| = \left| \frac{-\omega^2 LC}{1 - \omega^2 LC} \right|$$

If we normalize by using the natural frequency $\omega_N = 1/\sqrt{LC}$, then:

$$M = \frac{(\omega/\omega_N)^2}{1 - (\omega/\omega_N)^2}$$

Figure 8.23 shows the magnitude-frequency response for this circuit. We may observe at once that the circuit passes high frequency signals and blocks low frequency signals. Thus the characteristics are the same in this respect as a high-pass filter. There is a dissimilarity with an ideal high-pass filter, however, for applied signals near the natural frequency ($\omega/\omega_N=1$).

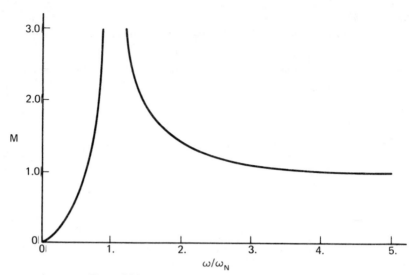

Figure 8.23 Magnitude-frequency curve (example 8.11)

For the circuit under consideration the output increases without bound in this vicinity. If the frequency band between ω/ω_N = 0.8 and ω/ω_N = 1.7 is never present in the input signal or can otherwise be avoided, the circuit performs reasonably well as a high-pass filter.

Band-Pass and Notch Filters

The combination of a low-pass and a high-pass filter results in a band-pass filter. Figure 8.24a shows the ideal band-pass filter characteristics. At very low and very high frequencies there is no output amplitude. Only frequencies in a band around some intermediate frequency produce an output signal. The ideal characteristics, of course, are just approximated very roughly by passive circuits. Active filters are required to approach the ideal characteristics more closely.

A low-pass and a high pass-filter can also be constructed to reject a specific frequency band and to pass all other frequencies. This type of filter is often called a notch filter. The characteristics of an ideal notch filter are

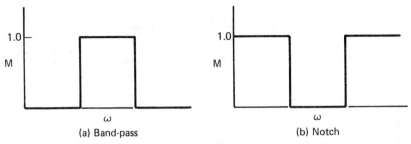

(a) Band-pass   (b) Notch

Figure 8.24 Ideal band-pass and notch filter characteristics

shown in Figure 8.24b. In this filter, the low and high frequency signals are passed. A narrow band in the intermediate frequency range is blocked. Here again, passive circuits only produce a crude approximation to the ideal characteristics. Sometimes this is sufficient for practical purposes. Many measuring instruments are disturbed by 60 Hz signals. A notch filter tuned to 60 Hz may be used to eliminate these undesirable signals. However, care must be exercised in filter selection to ensure that the filter does not also block desired frequencies.

The following examples demonstrate some of the crudeness of passive band-pass and notch filters.

Example 8.12

Figure 8.25 shows a bridged-T electric circuit. The circuit has two resistors and two capacitors. In this example, the resistors are equal and the capacitors are equal. (This is not necessary in general). The input signal is a sine wave. Investigate the magnitude-frequency characteristics of the circuit.

We may use the nodal circuit analysis to determine the transfer function for this circuit. The result is:

$$G(s) = \frac{E_O}{E_i} = \frac{(RC)^2 s^2 + 2(RC)s + 1}{(RC)^2 s^2 + 3(RC)s + 1}$$

The magnitude equals $|G(j\omega)|$ so that:

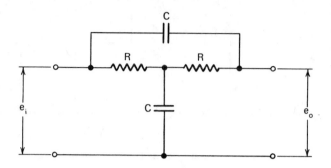

Figure 8.25 Bridged-T circuit (example 8.12)

$$M = \left| \frac{1 - (\omega RC)^2 + j(2\omega RC)}{1 - (\omega RC)^2 + j(3\omega RC)} \right| = \left[\frac{(1 - (\omega RC)^2)^2 + 4(\omega RC)^2}{(1 - (\omega RC)^2)^2 + 9(\omega RC)^2} \right]^{\frac{1}{2}}$$

Figure 8.26 shows the magnitude-frequency character-istics for the bridged-T circuit. At very low and very high frequency the input signal is not attenuated. In the vicinity of $\omega RC = 1$ the magnitude is reduced. How-ever, the circuit does not block any frequency complete-ly. The degree of blockage may be improved by using two different resistors or two different capacitors. We may consider the bridged-T circuit as a crude type of notch filter.

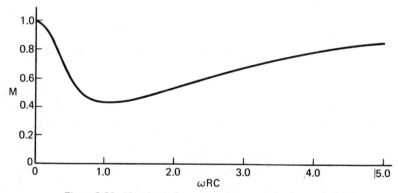

Figure 8.26 Magnitude-frequency characteristics (example 8.12)

Example 8.13

Investigate the magnitude-frequency characteristics of the electric LC circuit shown in Figure 8.27.

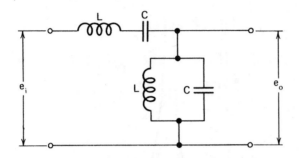

Fig. 8.27 Electric circuit (example 8.13)

The transfer function of the circuit is:

$$G(s) = \frac{LCs^2}{(LC)^2 s^4 + 3LCs^2 + 1}$$

and in the frequency domain:

$$G(j\omega) = \frac{-\omega^2 LC}{\omega^4 (LC)^2 - 3\omega^2 LC + 1}$$

If we define $\omega_N = 1/\sqrt{LC}$ the magnitude of the transfer function becomes:

$$M = \frac{(\omega/\omega_N)^2}{1 + (\omega/\omega_N)^4 - 3(\omega/\omega_N)^2}$$

The above equation is shown plotted in Figure 8.28. These characteristics indicate that the magnitude is zero at low and high frequencies. Thus the circuit is an effective block at the frequency extremes. However, at intermediate frequencies signals will pass. Unfortunately there are two frequencies ($\omega/\omega_N = .62$ and 1.62)

where the amplitude is unbounded. These frequencies
would have to be avoided. In general, we would classify
the circuit as a band-pass filter.

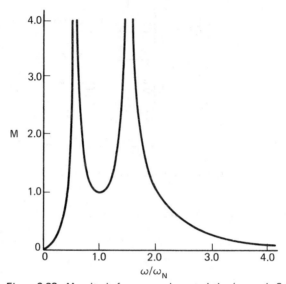

Figure 8.28   Magnitude-frequency characteristics (example 8.13)

8.5 RESONANCE

Resonance is a condition that occurs in circuits with
reactive or energy storage components when the frequency
response is maximum. If the transfer function, $G(j\omega)$,
represents an impedance or admittance, resonance comes
about when there is no phase shift. The condition for
resonance in such a circuit is therefore:

$$\underline{/G(j\omega_r)} = 0 \qquad\qquad (8.42)$$

where ω_r is the frequency which satisfies the condition
given in Equation (8.42) and is called the resonant frequency.

In many cases resonant circuits are deliberately sought
to perform a highly selective frequency discrimination.

These resonant circuits may be considered as a special type of band-pass filter, such as would be required in tuners for radio signals. On the other hand, however, resonance may be an extremely undesirable condition and we may try to avoid it. This would be the case in mechanical structural circuits where excessive strain could produce a catastrophic mechanical failure. In either case we should know how to analyze resonant circuits. Then we may design the circuit to use resonance or to avoid it.

There are basically two types of resonant circuits: (a) series resonance and (b) parallel resonance. We will discuss these circuits in turn.

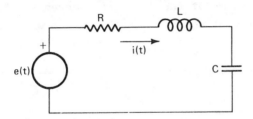

Figure 8.29 Series RLC electric circuit

Series Resonance

Figure 8.29 shows a series RLC electric circuit. The input voltage, e(t), is assumed to be a sine wave of the form E sinωt. The output signal is the current, i(t). The relation between output and input for expontential inputs is called the admittance, Y(s) and may be expressed for this circuit as:

$$Y(s) = \frac{I}{E} = \frac{Cs}{LCs^2 + RCs + 1} \qquad (8.43)$$

In the frequency domain Equation (8.43) becomes:

$$Y(j\omega) = \frac{j\omega C}{1 - \omega^2 LC + j(\omega RC)} \qquad (8.44)$$

and in terms of magnitude and phase:

$$Y(j\omega) = \frac{\omega C}{[(1-\omega^2 LC)^2 + (\omega RC)^2]^{\frac{1}{2}}} \Big/ 90° - \tan^{-1} \frac{\omega RC}{1-\omega^2 LC} \quad (8.45)$$

The resonant condition (Equation (8.42)) may now be applied to the phase portion of Equation (8.45). To obtain a zero phase the denominator of the inverse tangent function must be zero so that $1 - \omega_r^2 LC = 0$ and

$$\omega_r = 1/\sqrt{LC} \quad\quad (8.46)$$

Thus the resonant frequency is equivalent to the natural frequency of the circuit (ω_N) for the case of series resonance. In terms of the resonant frequency, the magnitude portion of Equation (8.45) may be written as:

$$M = |Y(j\omega)| = \frac{\frac{1}{R}\left[\frac{1}{Q}\frac{\omega}{\omega_r}\right]}{([1 - (\frac{\omega}{\omega_r})^2]^2 + [\frac{1}{Q}\frac{\omega}{\omega_r}]^2)^{\frac{1}{2}}} \quad (8.47)$$

where Q is a quality factor of resonator selectivity and has the form $\sqrt{L}/(R\sqrt{C})$. This Q factor is inversely related to the damping ratio of the circuit $(Q = 1/[2\delta])$. Equation (8.47) is shown plotted in Figure 8.30 for $Q = 1.25$ and 2.50. These values of Q were obtained by changing the value of R from 2 ohms to 1 ohm with L and C at fixed values. The magnitude in Figure 8.30 represents admittance. Thus we observe that for series resonance the driving point admittance is maximized and the driving point impedance is minimized.

For tuners Q values of about 20 000 are commonly used. This increases the selectivity of the circuit so that it, in effect, passes a very narrow band of frequencies.

A fluid circuit with the same analogous circuit diagram is known as the Helmholtz resonator (Figure 8.31). This device consists of a chamber with a short neck entrance passage. The R and L portion of the circuit are due to

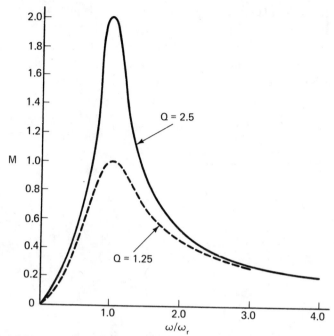

Figure 8.30 Effect of Q on magnitude-frequency characteristic of series RLC circuit

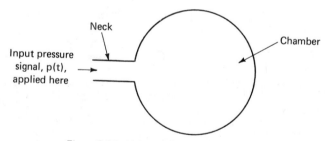

Figure 8.31 Helmholtz resonator

the neck. In this case, the R is really non-linear and we would have to use a linear approximation to analyze the circuit. The chamber, of course, represents the capacitance.

Parallel Resonance

Figure 8.32 shows an electric circuit which exhibits parallel resonance. The circuit has a capacitor in

parallel with a series RL combination. The input is the
current i(t) and the output is the voltage across the
capacitor e(t). In this circuit the relation between
output and input for exponential inputs is the driving
point impedance Z(s), and appears as:

$$Z(s) = \frac{E}{I} = \frac{R + Ls}{LCs^2 + RCs + 1} \tag{8.48}$$

and in the frequency domain:

$$Z(j\omega) = \frac{R + j\omega L}{1 - \omega^2 LC + j(\omega RC)} \tag{8.49a}$$

$$= \frac{R(1-\omega^2 LC) + \omega^2 RLC + j[\omega L-\omega^3 L^2 C-\omega R^2 C]}{[(1 - \omega^2 LC)^2 + (\omega RC)^2]} \tag{8.49b}$$

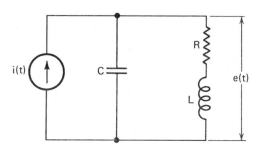

Figure 8.32 Electric circuit for parallel resonance

The resonant condition is that the reactive portion of
Equation (8.49b) must equal zero. Thus $(\omega_r L-\omega_r^3 L^2 C-\omega_r R^2 C) = 0$
and:

$$\omega_r = \frac{1}{\sqrt{LC}} \sqrt{1 - \frac{R^2 C}{L}} \tag{8.50a}$$

$$\omega_r = \omega_N \sqrt{1 - 1/Q^2} \tag{8.50b}$$

When R is small (high Q) the resonant frequency for
parallel and series resonance is the same. The magnitude

for the parallel resonant circuit, Equation (8.49b) may be expressed as:

$$M = \frac{R[1 + (Q \frac{\omega}{\omega_N} [1 - \frac{\omega^2}{\omega_N^2}] - \frac{1}{Q} \frac{\omega}{\omega_N})^2]^{\frac{1}{2}}}{[(1 - \frac{\omega}{\omega_N}]^2)^2 + [\frac{1}{Q} \frac{\omega}{\omega_N}]^2} \qquad (8.51)$$

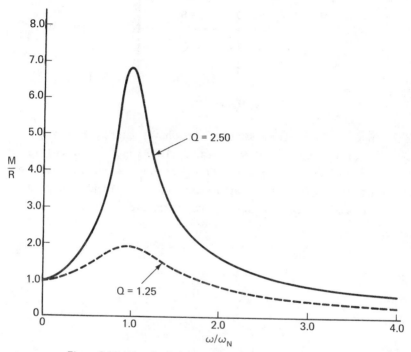

Figure 8.33 Magnitude-frequency response for parallel resonance

Figure 8.33 shows the magnitude-frequency response for parallel resonance. In this case the driving point impedance is maximum at resonance. Recall that for the series circuit the driving point impedance is minimum at resonance.

Problems

8.1 The input torque to the rotational flywheel-damper system shown in Figure P8.1 is described by

T(t) = T cosωt.   If the angular velocity of the
flywheel is considered as the output find the complete response of the system and the frequency
response.

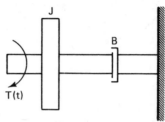

Figure P 8.1

8.2 In the electrical RC circuit shown in Figure P8.2
    R = 1 000 ohms and C = 5 μF.   The input is a voltage
    source e(t) = 100 cos 100 t, V and the output is the
    voltage across the capacitor, $e_c(t)$.   Find and plot
    the sinusoidal steady state response, $e_c(t)$.

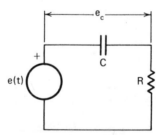

Figure P 8.2

8.3 A fluid line has a source of pressure connected to
    one end and is open to atmosphere at the other end.
    The pressure source may be described by p(t) = 40
    cos 0.1 t, Pa.   If the line may be modelled as
    R = $8(10)^5$ N·s/m^5 and L = $8(10)^6$ N·s^2/m^5 in series,
    find the frequency response for the flow in the line.

8.4 A plate of negligible mass is connected to a wall by
    a mechanical spring and viscous damper as shown in
    Figure P8.4.   The input force applied to the plate
    is f(t) = F sinωt and k = 18 N/m and b = 1.6 N·s/m.

If the velocity of the plate is considered as the output find the frequency at which the magnitude ratio M = 1/2. Also find the corresponding value of phase angle.

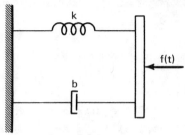

Figure P 8.4

8.5 In the rotational system shown in Figure P8.5 J = 4 kg·m², K = 20 N·m and B = 2 N·m·s. The torque input, T(t), is 10 sin 5t, N·m. Find the sinusoidal steady state response for the angular velocity of the flywheel.

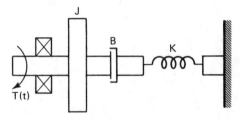

Figure P 8.5

8.6 In an RLC series electric circuit R = 200 ohms, L = 0.15 H, and C = 60 μF. The voltage source provides a signal with a frequency of 200 rad/s and an amplitude of 60 V. Find the sinusoidal steady state response of the current in the circuit.

8.7 In the mass-damper system shown in Figure P8.7 a velocity is applied to the free end of the damper so that v(t) = 100 sin 0.5t, m/s. The mass m=20 kg. The magnitude response for the velocity of the mass is M = 0.25. Find the value of the damping.

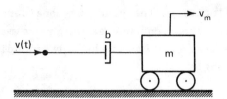

Figure P 8.7

8.8 The input impedance to many electronic measuring
instruments is often modelled as a resistance of
2 MΩ in parallel with a capacitance of 50 pF.
Determine the series equivalent when the frequency
is: (a) 15.92 Hz; (b) 1.592 kHz; (c) 159.2 kHz.

8.9 A damper of 2 N·s/m is connected in parallel with a
spring of 5 N/m. Find the series equivalent when
the frequency is: (a) .04 Hz; (b) .4 Hz; (c) 4 Hz

CHAPTER 9

MUTUALLY COUPLED COMPONENTS

Up to this point we have dealt exclusively with two-terminal components. These components were described by a relation between one through variable and one across variable difference. As a result each two terminal component was represented completely by a single impedance or admittance.

We now consider a class of components with two sets of signal variables. These components, which are generally referred to as two-terminal pairs, have two through variables and two across variables. Since there is a coupling between the variables of each pair and those of the other pair, it requires more than one impedance or one admittance to describe two-terminal pair networks.

9.1 TWO-TERMINAL PAIRS

A block diagram representation of a two-terminal pair network is shown in Figure 9.1. We designate the left side as terminal - pair 1 and the right side as terminal - pair 2. The network (within the block) may consist of combinations of conventional two-terminal components or may be a mutually coupled component that we will describe in Sections 9.2 and 9.3. However, whatever the implementation of the two-terminal pair its mathematical description has the same form.

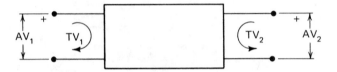

Figure 9.1 General two-terminal-pair network

To obtain the standard form for two-terminal pair networks we must recognize that an input (across or through variable type) at one terminal of the pair produces a response at that pair and also at the other pair. Suppose, for example, that the two-terminal pair is stimulated by through variable sources TV_1 and TV_2 (Figure 9.2a). Then by using superposition the responses of the corresponding across variables (AV_1 and AV_2) are:

$$AV_1 = Z_{11} \; TV_1 + Z_{12} \; TV_2 \qquad\qquad (9.1a)$$

$$AV_2 = Z_{21} \; TV_1 + Z_{22} \; TV_2 \qquad\qquad (9.1b)$$

where Z_{11} is the driving point impedance at terminal 1 under the condition that terminal 2 is open-circuited ($TV_2 = 0$). Similarly Z_{22} is the driving point impedance at terminal 2 with terminal 1 open-circuited ($TV_1 = 0$). The impedance Z_{12} represents the effect that source TV_2

has on AV_1 when $TV_1 = 0$. Correspondingly, Z_{21} relates
TV_1 and AV_2 when $TV_2 = 0$. In mathematical terms these
impedances may be written as:

$$Z_{11} = \left.\frac{AV_1}{TV_1}\right|_{TV_2 = 0} \qquad Z_{12} = \left.\frac{AV_1}{TV_2}\right|_{TV_1 = 0}$$

$$Z_{21} = \left.\frac{AV_2}{TV_1}\right|_{TV_2 = 0} \qquad Z_{22} = \left.\frac{AV_2}{TV_2}\right|_{TV_1 = 0}$$

(9.2)

Equations (9.1) and (9.2) indicate that we need four
impedances to describe two-terminal pair networks. However, from the reciprocity theorem for linear networks
the effect that TV_1 has on AV_2 is the same as the effect
that TV_2 has on AV_1. This means then that $Z_{12} = Z_{21}$ and
that we need only three impedances to describe the two-
terminal pair network.

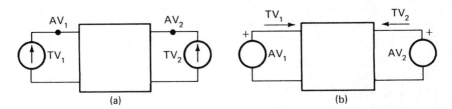

Figure 9.2 Two-terminal-pair networks

It is usually convenient to place the governing relationships for two-terminal pair networks in matrix form. Thus,
Equation (9.1) becomes:

$$\begin{bmatrix} AV_1 \\ AV_2 \end{bmatrix} = \begin{bmatrix} Z_{11} & Z_{12} \\ Z_{21} & Z_{22} \end{bmatrix} \begin{bmatrix} TV_1 \\ TV_2 \end{bmatrix}$$

(9.3)

Equation (9.3) is known as the impedance matrix representation of the two-terminal pair. There are other matrix
forms that can also be used to describe the performance of

the two-terminal pair. Consider the two-terminal pair
network with across variable sources AV_1 and AV_2 (Figure
9.2b). The responses of the corresponding through variables may be placed into the admittance matrix form:

$$\begin{bmatrix} TV_1 \\ TV_2 \end{bmatrix} = \begin{bmatrix} Y_{11} & Y_{12} \\ Y_{21} & Y_{22} \end{bmatrix} \begin{bmatrix} AV_1 \\ AV_2 \end{bmatrix} \tag{9.4}$$

where

$$Y_{11} = \left.\frac{TV_1}{AV_1}\right|_{AV_2 = 0} \qquad Y_{12} = \left.\frac{TV_1}{AV_2}\right|_{AV_1 = 0}$$

$$Y_{21} = \left.\frac{TV_2}{AV_1}\right|_{AV_2 = 0} \qquad Y_{22} = \left.\frac{TV_2}{AV_2}\right|_{AV_1 = 0}$$

and $Y_{12} = Y_{21}$ by reciprocity when the components are linear
and bilateral. The admittances in Equation (9.4), however,
are not the reciprocals of the impedances in Equation (9.3)
(e.g. $Y_{11} \neq Z_{11}$). The admittances are defined under short
circuit conditions (AV_1 or $AV_2 = 0$) whereas the impedances
are defined under open circuit conditions (TV_1 or $TV_2 = 0$).
This point was discussed in greater detail in Chapter VII.

   Another representation of the two-terminal pair network
is known as the hybrid matrix. This may be derived by
considering a through variable source at terminal 1 and
an across variable source at terminal 2. Thus:

$$\begin{bmatrix} AV_1 \\ TV_2 \end{bmatrix} = \begin{bmatrix} H_{11} & H_{12} \\ H_{21} & H_{22} \end{bmatrix} \begin{bmatrix} TV_1 \\ AV_2 \end{bmatrix} \tag{9.5}$$

where the values of H_{11}, H_{12}, H_{21} and H_{22} can be expressed
from Equations (9.3) or (9.4) in terms of driving point
and transfer impedances or admittances. The result is:

$$H_{11} = \frac{Z_{11}Z_{22} - Z_{12}Z_{21}}{Z_{22}} = \frac{1}{Y_{11}}$$

$$H_{12} = \frac{Z_{12}}{Z_{22}} = -\frac{Y_{12}}{Y_{11}}$$

$$H_{21} = -\frac{Z_{21}}{Z_{22}} = \frac{Y_{21}}{Y_{11}}$$

$$H_{22} = \frac{1}{Z_{22}} = \frac{Y_{11}Y_{22} - Y_{12}Y_{21}}{Y_{11}}$$

(9.6)

Equation (9.6) indicates that in the hybrid matrix $H_{12} = - H_{21}$ and also shows clearly that the elements in the impedance and admittance matrices are not reciprocals.

As stated previously the two-terminal pair components may be physically realized from a network of two-terminal components or from a mutually coupled component. The matrix representations given in Equations (9.3), (9.4) and (9.5) are valid for either physical realization. However, there is a reduced form of these matrices in which some of the matrix elements are equal to zero and which can only occur through the use of mutually coupled components. The reduced form is called an "ideal coupler". It occurs when the conditions $Z_{11}Z_{22} - Z_{12}Z_{21} = 0$ and $Y_{11}Y_{22} - Y_{12}Y_{21} = 0$ are fulfilled. Under these conditions the hybrid matrix elements H_{11} and H_{22} equal zero. Equation (9.5) then becomes the matrix representation of the ideal coupler and takes the form:

$$\begin{bmatrix} AV_1 \\ TV_2 \end{bmatrix} = \begin{bmatrix} 0 & n \\ -n & 0 \end{bmatrix} \begin{bmatrix} TV_1 \\ AV_2 \end{bmatrix}$$

(9.7)

where n is called the ideal coupling or transformation ratio.

Since our emphasis in this chapter is on ideal couplers we restrict our consideration to those two-terminal networks that are realized by mutually coupled components.

Specifically, we examine the most well-known mutually coupled element, the transformer, and show the conditions under which it approximates an ideal coupler.

9.2 TRANSFORMERS

Mutually Coupled Coils (Electrical Transformer)

Equation (3.7) expressed the relation that in a pure inductor (conducting coil) the flux linkage, λ_{12}, is directly proportional to the current. Faraday's Law was then used to develop the elemental equation of the inductor. This law states that the voltage induced in a coil is equal to the rate of change of flux linkage. Thus, the result developed in Equation (3.9) was that the induced coil voltage is proportional to the rate of change of current.

We may extend these basic concepts to the case where two separate coils are in close proximity to each other. In this arrangement the change of current in one of the coils induces voltage in both the coil itself and also in the nearby coil. These coils are therefore designated as "mutually coupled" and the phenomenon is known as mutual inductance.

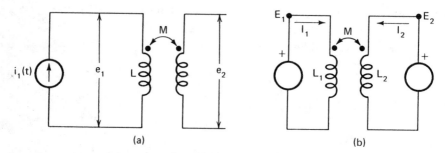

(a)   (b)

Figure 9.3 Mutually coupled coils

To describe the mutual inductance effect as a relation between signal variables consider the mutually coupled coils shown in Figure 9.3a. Suppose coils 1 and 2 are

mutually coupled and the current $i_1(t)$ flows in coil 1.
If we designate the voltage induced in coil 1 as e_1 and
the voltage induced in coil 2 as e_2 we may express these
voltages as:

$$e_1 = L \frac{di_1}{dt} \qquad (9.8a)$$

$$e_2 = M \frac{di_1}{dt} \qquad (9.8b)$$

where L is the self-inductance of coil 1 and M is the
mutual inductance between coil 1 and coil 2. The units
for mutual inductance and self inductance are the same
(henrys).

Figure 9.3b shows a two-terminal pair that consists of
mutually coupled coils and two voltage sources. The in-
clusion of M and the double-arrowed line indicates that
the coils are mutually coupled. The dots show the polarity. If current enters one coil at a dot, with a positive
rate of change, it induces a positive voltage on the dot
side of the other coil. If the loop equations (refer to
Section 7.5) are applied to this arrangement the result
is:

$$\text{loop } 1 \rightarrow -E_1 + L_1 s\, I_1 + Ms\, I_2 = 0 \qquad (9.9a)$$

$$\text{loop } 2 \rightarrow -E_2 + L_2 s\, I_2 + Ms\, I_1 = 0 \qquad (9.9b)$$

Equation (9.9) may be placed in the two-terminal pair
matrix forms given in Equations (9.3), (9.4) and (9.5).
The impedance matrix representation (Equation (9.3)) is:

$$\begin{bmatrix} E_1 \\ E_2 \end{bmatrix} = \begin{bmatrix} L_1 s & Ms \\ Ms & L_2 s \end{bmatrix} \begin{bmatrix} I_1 \\ I_2 \end{bmatrix} \qquad (9.10)$$

The corresponding admittance matrix (Equation (9.4)) for
the mutually coupled coils is:

$$
\begin{bmatrix} I_1 \\ \\ I_2 \end{bmatrix} = \begin{bmatrix} \dfrac{L_2}{(L_1 L_2 - M^2)s} & \dfrac{-M}{(L_1 L_2 - M^2)s} \\ \\ \dfrac{-M}{(L_1 L_2 - M^2)s} & \dfrac{L_1}{(L_1 L_2 - M^2)s} \end{bmatrix} \begin{bmatrix} E_1 \\ \\ E_2 \end{bmatrix} \tag{9.11}
$$

Note again that the corresponding elements in the impedance matrix (Equation (9.10)) and the admittance matrix (Equation (9.11) are not reciprocals. We may also put Equation (9.9) into the hybrid matrix form (Equation (9.5)) with the result:

$$
\begin{bmatrix} E_1 \\ \\ I_2 \end{bmatrix} = \begin{bmatrix} \dfrac{(L_1 L_2 - M^2)s}{L_2} & \dfrac{M}{L_2} \\ \\ -\dfrac{M}{L_2} & \dfrac{1}{L_2 s} \end{bmatrix} \begin{bmatrix} I_1 \\ \\ E_2 \end{bmatrix} \tag{9.12}
$$

Under special conditions the hybrid matrix for the mutually coupled coils (Equation (9.12)) may reduce to the form for the ideal coupler (Equation (9.7)). One of the conditions is that all the flux from one coil must be linked with the other coil. When this condition is met $M = \sqrt{L_1 L_2}$. The other condition is that virtually no current is required to produce the magnetic flux linking the coils. This means that L_1, L_2 and M are very large (approaching infinite values) and therefore $1/L_2$ approaches zero. However, the ratio M/L_2 is determinate and may be expressed as:

$$
\frac{M}{L_2} = \frac{\sqrt{L_1 L_2}}{L_2} = \sqrt{\frac{L_1}{L_2}} = \frac{N_1}{N_2}
$$

since in general the self-inductance of a coil is proportional to the square of the number of turns (e.g., see Equation (3.10)). If these conditions are satisfied, Equation (9.12) reduces to:

$$\begin{bmatrix} E_1 \\ I_2 \end{bmatrix} = \begin{bmatrix} 0 & N_1/N_2 \\ -N_1/N_2 & 0 \end{bmatrix} \begin{bmatrix} I_1 \\ E_2 \end{bmatrix} \tag{9.13}$$

Note that none of the terms in Equation (9.13) includes the complex frequency, s. However, it does not follow that in the sinusoidal steady-state (s = jω) the electrical transformer will behave approximately as an ideal trans- former at all frequencies. This becomes evident when we consider the term $1/j\omega L_2$ (H_{22} in Equation (9.12)) which clearly cannot be zero when ω = 0. Indeed, if we recall that when ω = 0 there is zero rate of change of the currents, we must conclude that the transformer cannot function under such a condition. Thus a pair of coupled coils may be represented by an ideal transformer only when the two conditions considered above are met. In general, these conditions cannot be realized, even approximately, at low frequencies.

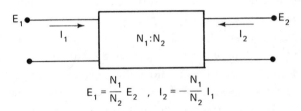

Figure 9.4 Representation of ideal electrical transformer

We use a block with the turns ratio (N_1 : N_2) indicated within as the circuit symbol for the ideal transformer. This is shown in Figure 9.4. For convenience, one side of a transformer is designated as "primary" and the other side as "secondary". The choice is somewhat arbitrary, but if there is only one source, it is common to consider as primary the side of the transformer connected to the source. For this discussion we shall arbitrarily specify

the left side as primary and the right side as secondary.
The number of turns on the primary is N_1 and in general,
the subscript 1 refers to the primary. Similarly N_2 is
the number of secondary turns, the subscript 2 being used
for all secondary quantities. The ideal transformer
relations may be simply expressed from Equation (9.13) as:

$$E_1 = (N_1/N_2)E_2 \qquad\qquad (9.14a)$$

$$I_2 = -(N_1/N_2)I_1 \qquad\qquad (9.14b)$$

Note that the voltage is always larger on the side having
the coil with the larger number of turns. The current is
smaller on this side. The ideal transformer neither stores
nor dissipates energy. It functions only to transform the
variables as described in Equation (9.14).

Mutually Coupled Mass (Mechanical Transformer)

Figure 9.5 shows an unpivoted bar of length, ℓ, and
mass, m. When forces F_1 and F_2 are applied at each end,
the bar has both translational and rotational motion. The
resulting velocity at the left end is v_1 and that at the
right end is v_2. The velocity at the center of gravity
of the bar is v_c. As in all the other components the
analysis must stem from physical laws. In this case we
use the basic laws of mechanics and then interpret the
results in a circuit context. For the translational
motion, $\Sigma f = $ (mass)(acceleration), and in terms of exponential amplitudes this may be expressed as:

$$F_1 + F_2 = ms\, V_c \qquad\qquad (9.15)$$

For the rotational motion,

$$\Sigma T = \text{(moment of inertia)(angular acceleration)}$$

This leads to:

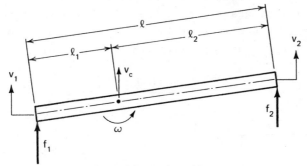

Figure 9.5 Unpivoted bar

$$F_2 \ell_2 - F_1 \ell_1 = J s \Omega \qquad (9.16)$$

where ℓ_1 and ℓ_2 = the distances from the ends to the center of gravity, Ω = the exponential amplitude of the angular velocity and J = the moment of inertia of the bar. To place Equations (9.15) and (9.16) into circuit format, we must separate the effects of the force at each end. The simultaneous solution of these equations yields:

$$m \ell_2 s V_C - J s \Omega = F_1 \ell \qquad (9.17a)$$

$$m \ell_1 s V_C + J s \Omega = F_2 \ell \qquad (9.17b)$$

Equation (9.17) represents the motion in terms of the angular velocity, Ω, and the velocity of the center of gravity, V_C. For a two-terminal pair circuit representation we require the motion to be in terms of V_1 and V_2, the velocities corresponding to the applied forces F_1 and F_2. We may relate these sets of variables (Ω, V_C and V_1, V_2) from the concept of relative motion. When the angle of rotation is small, $V_1 = V_C - \ell_1 \Omega$ and $V_2 = V_C + \ell_2 \Omega$ and thus:

$$V_C = \frac{V_1 \ell_2 + V_2 \ell_1}{\ell} \qquad (9.18a)$$

$$\Omega = \frac{V_2 - V_1}{\ell} \qquad (9.18b)$$

If the radius of gyration of the bar is a distance h from the center of gravity then the relation between the moment of inertia and mass is $J = mh^2$. The elimination of Ω and V_C from Equations (9.17) and (9.18) yields:

$$m \; \frac{(\ell_2^2 + h^2)}{\ell^2} \; s \; V_1 + m \; \frac{(\ell_1 \ell_2 - h^2)}{\ell^2} \; s \; V_2 = F_1 \qquad (9.19a)$$

$$m \; \frac{(\ell_1 \ell_2 - h^2)}{\ell^2} \; s \; V_1 + m \; \frac{(\ell_1^2 + h^2)}{\ell^2} \; s \; V_2 = F_2 \qquad (9.19b)$$

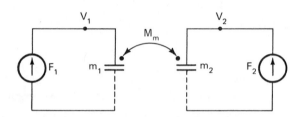

Figure 9.6 Mutually coupled masses

We may observe from Equation (9.19) that the relation between V_1 and F_1 also depends on V_2. Similarly, the relation between V_2 and F_2 also depends on V_1. From a circuit viewpoint, this is equivalent to a mutual coupling effect. Figure 9.6 shows a circuit representation of Equation (9.19) where M_m represents this mutual coupling effect. From the nodal equations we may analyze the circuit as:

$$\text{node } 1 \to m_1 s V_1 + M_m s V_2 = F_1 \qquad (9.20a)$$

$$\text{node } 2 \to M_m s V_1 + m_2 s V_2 = F_2 \qquad (9.20b)$$

Equation (9.20) is identical to Equation (9.19) when we define:

$$m_1 = m \; \frac{(\ell_2^2 + h^2)}{\ell^2} \qquad (9.21a)$$

$$m_2 = m \frac{(\ell_1^2 + h^2)}{\ell^2} \qquad (9.21b)$$

$$M_m = m \frac{(\ell_1 \ell_2 - h^2)}{\ell^2} \qquad (9.21c)$$

Equation (9.19) for the unpivoted bar may be placed in hybrid matrix form (Equation (9.12)) in terms of the analogous dual variables (i.e. F_1 in place of E_1 and V_2 in place of I_2, etc.) with the result:

$$\begin{bmatrix} F_1 \\ V_2 \end{bmatrix} = \begin{bmatrix} \dfrac{m\,h^2 s}{\ell^2} & \dfrac{\ell_1 \ell_2 - h^2}{\ell_1^2 + h^2} \\ -\dfrac{\ell_1 \ell_2 - h^2}{\ell_1^2 + h^2} & \dfrac{\ell^2}{ms(\ell_1^2 + h^2)} \end{bmatrix} \begin{bmatrix} V_1 \\ F_2 \end{bmatrix} \qquad (9.22)$$

Equation (9.22) represents the mutually coupled masses in the format of a two-terminal pair network. As in the case of the mutually coupled coils, we may apply special conditions to reduce the components to an ideal coupler. One of the conditions is that the mass is concentrated at the center of gravity ($h = 0$). The other condition is that the mass of the bar is large. With these conditions Equation (9.22) becomes:

$$\begin{bmatrix} F_1 \\ V_2 \end{bmatrix} = \begin{bmatrix} 0 & \ell_2/\ell_1 \\ -\ell_2/\ell_1 & 0 \end{bmatrix} \begin{bmatrix} V_1 \\ F_2 \end{bmatrix} \qquad (9.23)$$

The ideal unpivoted lever like the ideal electrical transformer requires a varying signal and does not operate down to zero frequency.

In dealing previously with two-terminal components, we have used the same circuit symbol in all the media. For example, mass and capacitance were represented similarly in circuit diagrams. We now extend the use of uniform symbols to the representation of ideal couplers. Thus, Figure 9.7 shows the ideal mechanical transformer with

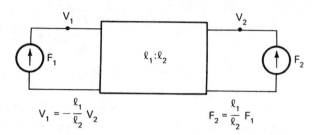

$$V_1 = -\frac{\ell_1}{\ell_2} V_2 \qquad\qquad F_2 = \frac{\ell_1}{\ell_2} F_1$$

Figure 9.7 Representation of ideal mechanical transformer

the same symbol that was used for the ideal electrical
transformer in Figure 9.4. In this case, the numbers
within the block (ℓ_1 : ℓ_2) refer to the distance from
the pivot point to the ends. The larger velocity occurs
on the side with the larger distance.

9.3 IDEAL COUPLERS

There are other physical arrangements which perform the
function of ideal couplers. These will be shown symbolically in the same way as in Figures 9.4 and 9.7. In all
cases the larger number within the block is placed on
the side with the larger value of the across variable.
We emphasize this now because some of the couplers operate
with the dual variable. As a result, a geometric parameter associated with one side may appear within the block
on the opposite side. This will become clear when we
describe the operation of gears and differential pistons.

Levers

In Section 9.2 we demonstrated that an unpivoted bar
with a large mass concentrated at its center of gravity
acts as an ideal coupler. We can obtain another ideal
coupling component by merely pivoting the bar about a
fixed support and assuming that the mass of the bar is
negligible. The component (Figure 9.8) is then an ideal
lever.

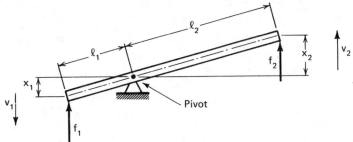

Figure 9.8 Pivoted lever

There are two relations that must be derived; one between v_1 and v_2 and the other between f_1 and f_2. We may relate the velocities from the geometry of the lever. For small angular displacements the motion of the ends is approximately linear and thus:

$$\frac{x_1}{\ell_1} = \frac{x_2}{\ell_2} \qquad (9.24)$$

since the lever arms ℓ_1 and ℓ_2 are fixed, we may differentiate Equation (9.24) with respect to time and obtain:

$$v_1 = (\ell_1/\ell_2)v_2 \qquad (9.25)$$

It is important to notice that the directions of v_1 and v_2 are selected to eliminate negative signs in the relation between the velocities. Inspection of any particular lever arrangement will always indicate the proper choice of direction to avoid negatives.

The relation between f_1 and f_2 may be accomplished by equating the energy input on one side to the energy output of the opposite side. However, for a lever without appreciable mass the moment of inertia is negligible. As a result the sum of the moments about the pivot are approximately equal to zero and we may obtain directly that:

$$f_2 = (\ell_1/\ell_2)f_1 \qquad (9.26)$$

Equations (9.25) and (9.26) are the relations between the primary and secondary variables for the ideal lever. The transformation ratio is ℓ_1/ℓ_2. Note also that the pivoted ideal lever operates at all frequencies including zero frequency.

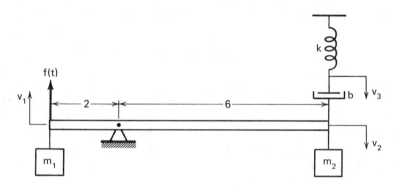

Figure 9.9 Mechanical arrangement for example 9.1

Example 9.1

A mechanical arrangement which includes an ideal lever, two masses, a spring and a damper is shown in Figure 9.9. The source is a vertical force f(t) applied at the left end of the lever. Draw the analogous circuit diagram for this system. In addition find the transformation ratio and the relation between the primary and secondary variables.

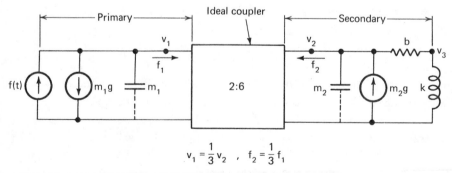

$$v_1 = \frac{1}{3}v_2 \quad , \quad f_2 = \frac{1}{3}f_1$$

Figure 9.10 Analogous circuit for example 9.1

Figure 9.10 shows the analogous circuit diagram. The primary and secondary sides are connected through an ideal coupler with the ratio 2:6. Since the velocity on the secondary side is larger than that on the primary side, the larger number appears on the secondary side. The primary circuit consists of the source due to the applied force, the force source due to gravity and the mass, m_1. The secondary circuit has a mass m_2, a damper b, a spring k, and a force source due to gravity acting on m_2.

The transformation ratio is 1/3 and the relations between the primary and secondary variables are:

$$v_1 = (1/3)v_2 \text{ and } f_2 = (1/3)f_1$$

Pulleys

Smooth pulleys of negligible mass may also be considered as ideal couplers. In this case, the transformation ratio depends on the number of lines supporting the pulley. The following examples demonstrate the determination of the analogous circuit for mechanical arrangements involving pulleys.

Example 9.2

In the pulley arrangement shown in Figure 9.11a, a force $f_1(t)$ is applied to the free end of a line that passes over a fixed pulley and under a movable pulley. The velocity of the free end is designated as v_1 in the direction of the force. The velocity of the movable pulley is v_2 and is assumed upward. Represent this mechanical device by an analogous circuit diagram if the pulleys are assumed to be smooth. Find the transformation ratio and the relation between primary and secondary variables.

Figure 9.11b shows a free body diagram of the movable pulley. The assumed smoothness of the pulley surfaces means that friction may be neglected. As a consequence,

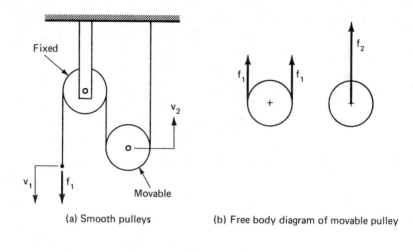

(a) Smooth pulleys   (b) Free body diagram of movable pulley

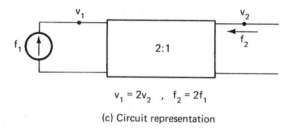

$$v_1 = 2v_2 \quad , \quad f_2 = 2f_1$$

(c) Circuit representation

Figure 9.11   Pulley arrangement for example 9.2

the applied tensile force, f_1, is transmitted equally throughout the line. Thus because of the loop in the line the movable pulley is acted upon by two equal and parallel forces, f_1. The resultant force acting on the movable pulley, f_2, is therefore equal to $2 f_1$. Geometric considerations show that the movable pulley moves only half the distance of the free end. Thus the transformation ratio is 2 and the relation between the variables is $v_1 = 2v_2$, $f_2 = 2f_1$.

   Figure 9.11c shows the circuit representation of the pulley arrangement. The pulley behaves as an ideal coupler. Note that the number of lines supporting the movable pulley which is associated with the secondary variables is placed within the block on the primary side (2:1).

Example 9.3

Figure 9.12a shows a hoist that has a force, $f(t)$, applied through a spring, k. The velocity of the free end of the hoist cable is v_1. The hoist has a fixed set of two smooth pulleys and a movable set of two smooth pulleys. The velocity associated with the movable set is v_2. The object of the device is to lift the mass, m. Represent the hoist by an analogous circuit. Determine the transformation ratio and the relation between the variables on the primary and secondary sides.

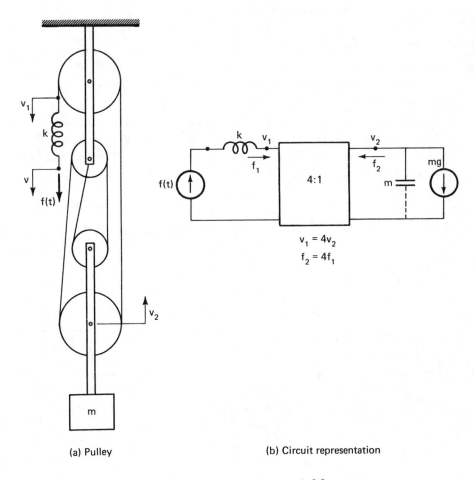

(a) Pulley   (b) Circuit representation

Figure 9.12 Pulley arrangement for example 9.3

The tension in the hoist cable, f_1, is equal to $f(t)$. The cable is looped in such a way that the resultant force, f_2, acting on the movable pulleys is 4 times the tensile force. Similarly, the motion of the free end of the cable is 4 times larger than that of the movable pulleys. Thus we may write $f_2 = 4f_1$, $v_1 = 4v_2$. These are the coupling equations of a device with a transformation ratio of 4.

Figure 9.12b shows the circuit representation of the hoist. The primary side has a spring in series with the force source. The secondary side has a gravity force source and a mass to ground.

Gears

In translational mechanical systems levers and pulleys are ideal couplers. The corresponding component in rotational mechanical systems is a set of gears. Figure 9.13a shows a simple arrangement of two gears. The number of teeth on gear 1 is N_1 and on gear 2 is N_2. The relation between the angular velocity of shafts 1 and 2 and the number of gear teeth is:

$$\omega_1 = (N_2/N_1)\omega_2 \qquad\qquad (9.27)$$

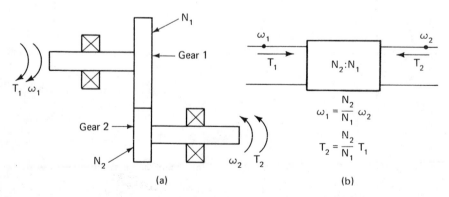

Figure 9.13 Representation of gears as an ideal coupler

The directions have been selected to avoid negatives. Comparison of Equation (9.27) with the equivalent relation for levers (Equation (9.25)) shows that the transformation ratio for the gear is N_2/N_1 whereas the ratio for the lever is ℓ_1/ℓ_2. To retain the convention that the larger number in the ideal coupling block be on the side with the larger across variable, we put N_2 on the primary side and N_1 on the secondary side. This is shown in Figure 9.13b. The relation between the torques associated with gears 1 and 2 is:

$$T_2 = (N_2/N_1)T_1 \qquad (9.28)$$

The gears, like the pivoted ideal lever, operate at all frequencies including zero frequency.

Example 9.4

In the gear arrangement shown in Figure 9.14 a 40 tooth gear drives an 80 tooth gear through a 20 tooth idler gear. An input torque, $T_1(t)$, is applied to the driving gear. The driven gear is connected to a large flywheel, J, and a damper, B. Represent the gear arrangement by an analogous circuit and determine the transformation ratio.

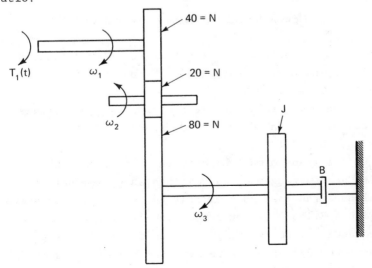

Figure 9.14 Gear arrangement for example 9.4

Let us designate the angular velocities of the driving, idler, and driven gears as ω_1, ω_2 and ω_3 respectively. There is an ideal coupling action between each pair of gears (i.e., 1 and 2, 2 and 3). Thus we may describe the rotational system with two ideal couplers, as shown in Figure 9.15a. The transformation ratio for coupler 1 (between gears 1 and 2) is 20:40 and so we may write that $\omega_1 = (1/2)\omega_2$, $T_2 = (1/2)T_1$.

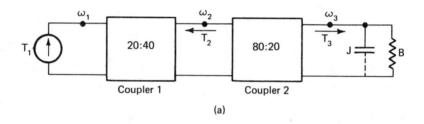

(a)

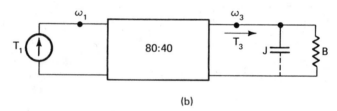

(b)

Figure 9.15 Circuit representation for example 9.4

The transformation ratio for ideal coupler 2 (between the idler and driven gear) is 80:20 and therefore $\omega_2 = 4\omega_3$, $T_3 = 4T_2$.

From the above relations we may express the variables associated with the driving and driven gears as:
$\omega_1 = 2\omega_3$, $T_3 = 2T_1$.

Thus the transformation ratio that represents the entire gear arrangement is 2. This suggests that we may reduce the analogous circuit to one coupler as shown in Figure 9.15b. Of course, we could have done this immediately by recognizing that the idler gear merely changes the direction of rotation of the driven shaft and does not affect its angular velocity or torque.

Differential Piston

The ideal coupler component for fluid systems is the differential piston shown in Figure 9.16a. On the primary side the pressure is p_1 and the flow q_1. On the secondary side the pressure is p_2 and the flow q_2. Between the two sides is an enclosed region that is connected directly to a fixed reference pressure. The most convenient reference pressure and the one most easily maintained is the atmospheric reference at pressure, p_g. If the piston has negligible mass the sum of the forces applied to it must equal zero and therefore:

$$(P_1 - P_g) = (A_2/A_1)(P_2 - P_g) \qquad (9.29)$$

where A_1 and A_2 are the cross-sectional areas on the primary and secondary sides respectively. Since there is only one piston velocity we may relate the primary and secondary flows by:

$$q_2 = (A_2/A_1)q_1 \qquad (9.30)$$

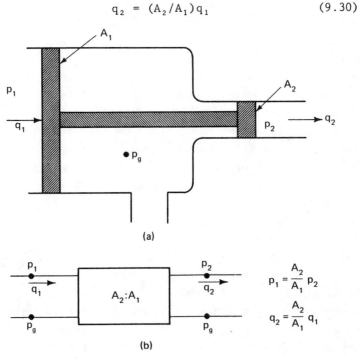

(a)

(b)

Figure 9.16 Differential piston

Equations (9.29) and (9.30) show that the differential
piston has the relationships required for an ideal coupler.
The circuit representation is presented in Figure 9.16b.
The transformation ratio is A_2/A_1. As in the gear and
pulley coupling the geometrical parameters are transposed
within the couplign block. That is the area A_2 appears
on the primary side and the area A_1 on the secondary side.
In this case it is the pressure which is always larger
on the side of the larger number.

9.4 EQUIVALENT NETWORK FOR IDEAL COUPLERS

In general the relations between the primary and secondary variables for ideal couplers may be expressed as:

$$(AV)_P = n \ (AV)_S \qquad (9.31a)$$

$$(TV)_S = n \ (TV)_P \qquad (9.31b)$$

where the subscripts P and s refer to primary and secondary. We may consider the transformation ratio, n, as a
scale factor between the sides. Thus, for example, it is
possible to place scaled secondary variables into a single
equivalent circuit with primary variables and to analyze
the resulting circuit by the methods we have previously
studied in Chapter VII. This procedure is sometimes
called "reflecting" or "referring" to the primary. Conversely we may also place scaled primary variables into a
single equivalent circuit with secondary variables. This
is "reflecting" or "referring" to the secondary.
When an ideally coupled circuit is put into the form of
a single equivalent circuit all the basic components,
sources, and signal variables must be scaled. Equation
(9.31) will help us to determine the required scaling of
sources and signal variables. Thus to reflect an across
variable or across variable source from secondary to
primary we must multiply by the transformation ratio. A

convenient way to remember this is to apply Equation
(9.31a) and to express it with the variable on the side
we are reflecting to by itself. The coefficient of the
other side is the scaling factor. To reflect from primary
to secondary then we would write Equation (9.31a) as
$(AV)_S = (1/n)(AV)_P$, and this would indicate division
by the transformation ratio.

For the case of a through variable or through variable
source we would apply the same procedure to Equation
(9.31b). To shift the through variable from secondary
to primary we would write Equation (9.31b) as
$(TV)_P = (1/n)(TV)_S$ and this would indicate division by
the transformation ratio. On the other hand, a shifting
to the secondary would mean using Equation (9.31b) in
its given form and then we would have to multiply by
the transformation ratio.

The scaling factors for the reflection of the basic
components are most conveniently treated by expressing
the components in terms of impedance or admittance. We
may define these in the primary and secondary circuits
as:

$$Z_P = \frac{(AV)_P}{(TV)_P} \quad , \quad Y_P = \frac{(TV)_P}{(AV)_P} \qquad (9.32a)$$

$$Z_S = \frac{(AV)_S}{(TV)_S} \quad , \quad Y_S = \frac{(TV)_S}{(AV)_S} \qquad (9.32b)$$

We may now relate Z_P to Z_S and Y_P to Y_S by using
Equation (9.31) in conjunction with Equation (9.32).
The result is:

$$Z_P = n^2 Z_S \qquad (9.33a)$$

$$Y_P = (1/n^2)Y_S \qquad (9.33b)$$

Equation (9.33a) may be interpreted in a similar manner
to Equation (9.31). When impedances are reflected from
secondary to primary we must multiply them by n^2. On

the other hand, to reflect impedances from primary to
secondary divide the impedances by n^2. To reflect admittances we should refer to Equation (9.33b). Thus admittances reflected from secondary to primary should be
divided by n^2 and those reflected from primary to secondary should be multiplied by n^2. The following examples
will clarify the procedure.

Example 9.5

Figure 9.17 shows the analogous circuit of a mechanical
system that includes an ideal coupler. The primary circuit consists of a velocity source, V_1, and a series
impedance Z_1. The secondary circuit has a series impedance, Z_2, a shunt impedance, Z_3, and a force source,
F_1. The transformation ratio is 3. Determine a single
equivalent circuit: (a) in terms of primary variables
and (b) in terms of secondary variables.

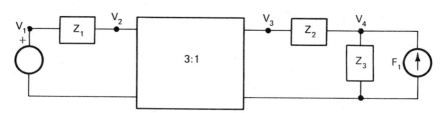

Figure 9.17 Analogous mechanical circuit (example 9.5)

For each circuit express the velocity V_4 in terms of
the sources and impedances.

(a) Primary Equivalent Circuit

. To obtain the primary equivalent circuit we must
multiply all secondary impedances by 9 (Equation (9.33a)).
The force source in the secondary must be divided by 3
(Equation (9.31b)) and the velocity V_4 must be multiplied
by 3 (Equation (9.31a)).

If there had been a velocity source in the secondary
we would have treated it similar to a velocity and

multiplied it by 3. The equivalent circuit in terms of the primary variables is shown in Figure 9.18. We may determine the variable, $3V_4$, by superposition. Thus:

$$\left[\frac{3V_4}{V_1}\right]_{F_1=0} = \frac{9Z_3}{Z_1 + 9Z_2 + 9Z_3}$$

$$\left[\frac{3V_4}{F_1/3}\right]_{V_1=0} = \frac{(9Z_3)(Z_1 + 9Z_2)}{Z_1 + 9Z_2 + 9Z_3}$$

The combined effect of both sources is then:

$$3V_4 = \frac{9Z_3(V_1)}{Z_1 + 9Z_2 + 9Z_3} + \frac{9Z_3(Z_1 + 9Z_2)(F_1/3)}{Z_1 + 9Z_2 + 9Z_3}$$

or

$$V_4 = \frac{3Z_3 V_1 + Z_3(Z_1 + 9Z_2)F_1}{Z_1 + 9Z_2 + 9Z_3}$$

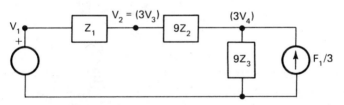

Figure 9.18 Equivalent network in terms of primary variables (example 9.5)

(b) Secondary Equivalent Circuit

To obtain the secondary equivalent circuit the impedance in the primary must be divided by 9 (Equation (9.33a)). The velocity source in the primary is divided by 3 when it is placed in the secondary (Equation (9.31a)). Had there also been a force source in the primary it would have required multiplication by 3 to bring it into the equivalent secondary circuit. Figure 9.19 shows the secondary equivalent.

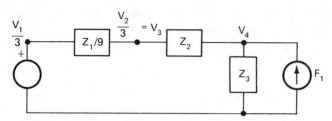

Figure 9.19 Equivalent network in terms of secondary variables (example 9.5)

Once again, we must use superposition to determine, V_4.

$$\left[\frac{V_4}{V_1/3}\right]_{F_1=0} = \frac{9Z_3}{Z_1 + 9Z_2 + 9Z_3}$$

$$\left[\frac{V_4}{F_1}\right]_{V_1=0} = \frac{Z_3(Z_1 + 9Z_2)}{Z_1 + 9Z_2 + 9Z_3}$$

and the combined effect is:

$$V_4 = \frac{3Z_3 V_1 + Z_3(Z_1 + 9Z_2)F_1}{Z_1 + 9Z_2 + 9Z_3}$$

Note that the results for V_4 are the same in both cases. We may, therefore, always use either the primary or secondary equivalent circuit to determine the value of any signal variable.

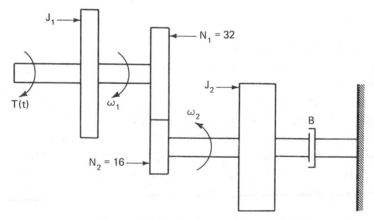

Figure 9.20 Mechanical circuit (example 9.6)

Example 9.6

A motor drives a flywheel through a gearing arrangement as shown in Figure 9.20. The motor has no friction
and has a moment of inertia J_1 = 1 000 kg·m². The
driving gear has 32 teeth and the driven gear has 16
teeth. The flywheel has a moment of inertia
J_2 = 2 250 kg·m² and flywheel friction may be modelled
as a damper with B = 250 N·m·s/rad. If the motor is
suddenly turned on and then develops a torque
T(t) = 2 000 u_S(t), N·m, find the angular velocity of
the flywheel, ω_2(t). The flywheel is initially at rest.

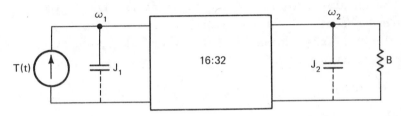

Figure 9.21 Analogous circuit (example 9.6)

Figure 9.21 shows the analogous circuit diagram. The
gears are represented by the ideal coupler with the
larger number on the side with larger angular velocity.
Thus, for this coupler the transformation ratio is 1/2.
To derive the differential equation for this circuit we
shall use the admittances of the components. This
requires the assumption of exponential signals through-
out the circuit (i.e. T(t) = T e^{st}, ω_2(t) = Ω_2 e^{st}).

We may proceed to obtain the governing equation for
this system using these exponential functions even though
the actual input torque is not an exponential function.

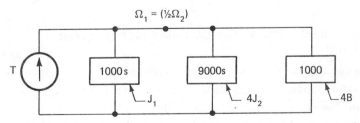

Figure 9.22 Equivalent circuit with primary variables and admittances (example 9.6)

For this system we express the components in terms of
admittances and reflect the secondary components to the
primary side as shown by the primary equivalent circuit
in Figure 9.22. When reflecting admittances from secondary to primary we must divide by n^2 as indicated in
Equation (9.33b). In this case $n = 1/2$ so that in
effect we must multiply the admittances by 4. Thus,
for the flywheel and damper in primary terms

$$Y_J = 4(2\ 250)s = 9\ 000\ s, \quad Y_B = 4(250) = 1\ 000$$

The angular velocity of the flywheel is $(1/2)\Omega_2$ in
accordance with Equation (9.31a). From a circuit
analysis of the equivalent primary circuit the driving
point admittance is:

$$Y_{dp} = \frac{T}{(1/2)\Omega_2} = 10\ 000\ s + 1\ 000$$

We may use this system function to find the differential equation in the time domain as:

$$10\ \frac{d[(1/2)]\omega_2}{dt} + (1/2)\omega_2 = \frac{T(t)}{1\ 000} = 2u_s(t)$$

The solution of this equation is:

$$(1/2)\omega_2(t) = 2(1 - e^{-t/10})$$

so that:

$$\omega_2(t) = 4[1 - e^{-t/10}]$$

9.5 SELECTION OF COUPLER FOR MAXIMUM POWER TRANSFER

We may transfer maximum power from a given source with
fixed impedance to a specific load by inserting an ideal
coupler between them. The procedure involves selecting

the transformation ratio so that the equivalent circuit meets the maximum power condition given in Equation (8.37).

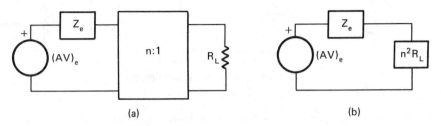

(a)   (b)

Figure 9.23 Transfer of power through an ideal coupler

Figure 9.23a shows the Thevenin circuit with an ideal coupler placed between the source and the resistive load. If we reflect the load circuit into an equivalent source circuit (Figure 9.23b) the value of the resistive load must be multiplied by n^2. If we now apply the condition for maximum power transfer (Equation (8.37)) the transformation ratio to achieve this is:

$$n_m^2 = \frac{\sqrt{R_e^2 + X_e^2}}{R_L} \tag{9.34}$$

where n_m is the transformation ratio for maximum power transfer, R_e is the real part and X_e the imaginary part of the impedance $Z_e(j\omega)$.

When the load is reactive it will generally not be possible to select an ideal coupler for maximum power transfer. In this case we would have to make some compromise between the two conditions given in Equation (8.39).

Example 9.7

An electric circuit containing an ideal transformer (Figure 9.24) has $R_1 = 7.5\ \Omega$, $L = 5\ H$, $C = 0.1\ F$, $R_L = 1\ \Omega$ and $e(t) = 10 \sin 2t$. Find the transformation ratio of the ideal transformer so that maximum power will be transferred to the 1 ohm resistor.

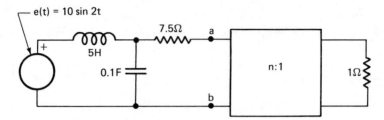

Figure 9.24 Electric circuit (example 9.7)

From the terminals ab on the primary side the equivalent impedance looking toward the source is:

$$Z_e(s) = R_1 + \frac{(Ls)(1/Cs)}{Ls + (1/Cs)} = \frac{R_1 LCs^2 + Ls + R_1}{LCs^2 + 1}$$

$$Z_e(j\omega) = \frac{R_1(1-\omega^2 LC) + j\omega L}{1 - \omega^2 LC} = 7.5 - j10$$

Now the application of Equation (9.34) yields:

$$n_m^2 = \frac{\sqrt{(7.5)^2 + (10)^2}}{1} = 12.5$$

$$n_m = 3.535$$

and the primary requires 3.535 more turns on its coil
than on the secondary coil. A transformer with transformation ratio equal to 3.535 will permit the maximum
power to be transferred to the load.

Problems

9.1 For the electrical circuit shown in Figure P9.1, the
transformer has a primary-to-secondary turns ratio
of 4. The primary has a capacitive reactance of
8 ohms, a resistance of 3 ohms and an inductive
reactance of 4 ohms. The rms value of the voltage
generator is 10 V. The secondary has a resistance
of 1 ohm. Determine the power dissipated in the

1 ohm resistor. Also find the current through and voltage drop across the 1 ohm resistor.

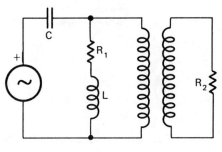

Figure P 9.1

9.2 Determine the output voltage, $e_o(t)$, for the electrical circuit shown in Figure P9.2. The source $e(t) = 20 \cos 2\,000\ t$, V and the turns ratio between primary and secondary is 2.

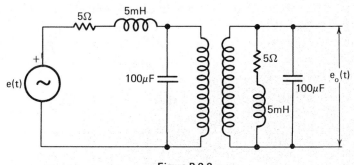

Figure P 9.2

9.3 A source has the equivalent circuit shown in Figure P9.3. The source supplies power to a load resistance of 8 ohms. To dissipate maximum power in this load resistance it is proposed to use a transformer which may be assumed to be ideal. Determine the turns ratio of the transformer which will result in maximum power being dissipated in the 8 ohm load. For this transformer calculate the power transferred

to the load when the voltage source has an rms value
of 20 V.

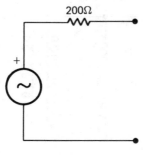

Figure P 9.3

9.4 A rotational mechanical system may be modelled by
the circuit shown in Figure P9.4. The system is
used to drive a load which is equivalent to a damper
of 1 N·s/rad.
   (a) If the drive system is connected directly to the
       load calculate the power transferred to the load.
   (b) It is desired to transfer the maximum power
       possible from the system to the load. Find the
       gear ratio which must be used to accomplish this.
       In addition, find the value of the maximum power
       and the resulting angular speed of the load shaft.

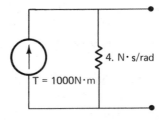

Figure P 9.4

9.5 An electric generator has a constant source voltage
and frequency. The equivalent circuit of the
generator is shown in Figure P9.5. When the gener-

ator is connected to a 20 ohm resistive load through
an ideal transformer find the turns ratio that
permits the maximum power to be transferred to the
load. Determine also the maximum power and the
voltage across the 20 ohm resistor.

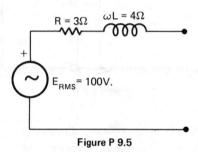

Figure P 9.5

9.6 A load resistance of 8 ohms is to be connected to
the circuit shown in Figure P9.6 so that maximum
power is dissipated in the resistance.
(a) Find the turns ratio of an ideal transformer
to achieve this condition.
(b) When maximum power is transferred what is the
voltage across the 8 ohm resistor?

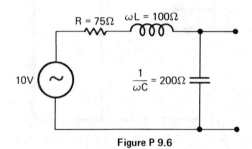

Figure P 9.6

9.7 A mechanical system contains a mass, spring,
damper and lever as shown in Figure P9.7. Determine the damping ratio of the system.

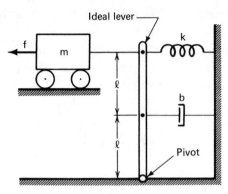

Figure P 9.7

9.8 The mechanical arrangement shown in Figure P9.8
contains a mass, spring, damper and ideal lever.
Find the position on the lever (x) at which the
damper must be connected so that the resulting
system will be critically damped. The mass
m = 1 kg, the spring, k = 4 N/m, the damper,
b = 9 N·s/m and the lever is 0.5 m in length.

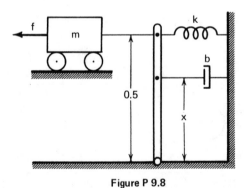

Figure P 9.8

9.9 A mass, spring, damper and lever arrangement is
shown in Figure P9.9. The mass is connected to
the end of the lever. The damper and spring are
connected to the lever at the points shown.

(a) Find the equivalent primary circuit for this arrangement.

(b) Determine the damping ratio of the system.

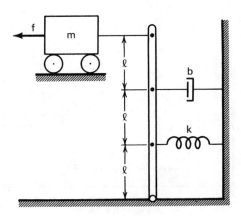

Figure P 9.9

9.10 A hydraulic press is represented in Figure P9.10. The cross-sectional area of the smaller cylinder is A_1 and that of the larger cylinder is A_2. Assuming that the fluid is incompressible and its mass is negligible represent the device as an ideal transformer.

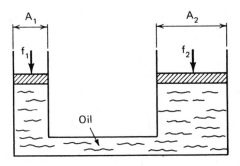

Figure P 9.10

9.11 A hydraulic press (Figure P9.11) with $A_1 = 0.5$ m^2 is used to lift an automobile of mass 1 000 kg.

(a) Find the force required at the input, f.

(b) Find the equivalent mass referred to the input.

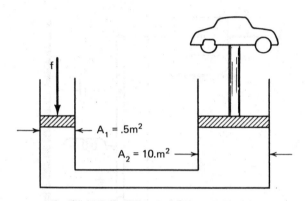

Figure P 9.11

ANSWERS
TO SELECTED
PROBLEMS

CHAPTER II

2.1 $y = \frac{2}{5} u_p(t) - \frac{4}{5} u_p(t-5) + \frac{2}{5} u_p(t-10)$

2.7 $g(t) = 5t^2 - 20t + 20, \quad t \geq 2$

$\qquad\quad = 0 \qquad\qquad\qquad\quad t \leq 2$

2.11 a) $g(t) = u_r(t) - 1.5u_r(t-1) - 1.5u_r(t-2) + 3u_r(t-3)$

$\quad$ c) $h(t) = 2u_s(t-1) - 2u_s(t-3)$

2.17 a) $g(t) = 2au_s(t) - 3au_s(t-T_1)$

$$+ \frac{a}{T_3-T_2} u_r(t-T_2) - \frac{a}{T_3-T_2} u_r(t-T_3)$$

2.23 2780 m

CHAPTER III

3.1   a) Capacitance, $C = 0.005$ F

      b) Inertance, $L = 5\,000$ N·s^2/m^5

      c) Capacitance, $C = 2.5(10)^{-10}$ m^5/N

      d) Resistance, $R = 250$ ohms

3.3   a) Damper (Resistance) $B = 0.6$ N·m·s/rad

      b) Spring (Inductance) $k = 800$ N/m

      c) Capacitance $C = 40(10)^{-10}$ m^5/N

      d) Mass (Capacitance) $m = 2.25$ kg

3.5   a) $F = 50 \sin 4t\ u_s(t)$, N

      b) $m = 5$ kg

3.7 $i(3.36) = 11.22$ A, $W(3.36) = 125.89$ joules

3.11 $W(5) = 0.045$ joules

3.13   a) $\ell = 4.58$ m, $d = 0.0985$ m

       b) $t = 19.53$ s

       c) $W = 0.00733$ joules

3.15 $q_o = 5$ m^3/s, $h = 3$ m

3.17   25 percent decrease in current

CHAPTER IV

4.1   a) 5 A,   0 V,   3.57 A,   7.14 V

      b) 0 V/s,   100 A/s

4.3 a) 0 rad/s, 100 rad/s, 500 N·m

b) 5 rad/s^2

4.11 a) 6 N/s, 1.6 N/s

b) 6.67 N, 1.67 m/s

CHAPTER V

5.1 a) t = 0.96 s   b) h = 8.34 m

5.3 a) $v_m(0.25)$ = 0.6, m/s, b) t = 1.89 s c) no effect

5.5 a) 131.25 W   b) τ = 38.7 hours

5.7 a) p(t) = 49 000 $e^{-t/40}$, Pa   b) h = 2.32 m

5.9 a) circuit 1 → $e_1(t)$ = $4e^{-t/RC}$, i(t) = $\frac{4}{R}$ $e^{-t/RC}$

circuit 2 → $e_1(t)$ = $4(1-e^{-t/RC})$, i(t) = $\frac{4}{R}$ $e^{-t/RC}$

b) circuit 1 → $e_1(t)$ = $\frac{20}{C}$ $u_r(t)$, i(t) = $20u_s(t)$

circuit 2 → $e_1(t)$ = $20Ru_s(t)$, i(t) = $20u_s(t)$

5.11 $\omega_1(t)$ = $[5t - \frac{5}{2}(1-e^{-2t})]u_s(t)$

$-[5(t-10) - \frac{5}{2}(1-e^{-2(t-10)})]u_s(t-10)$, rad/s

5.13 J = $1.035(10)^5$ kg·m^2

5.15 x = 4 m

5.17 $\omega(t)$ = $\frac{T}{B}$ - $[\frac{T}{B} + \omega_o]e^{-Bt/J}$

5.19 v(t) = 0.69 $e^{-.31t}$, m/s, x(t) = $0.69(1-e^{-.31t})$,

5.21 a) 61 313 Pa   b) 61 313 Pa

5.23 v = 1.47 m/s,   x = 2.72 m

CHAPTER VI

6.1 $v(t) = 0.5 \cos 4t$, m/s

6.3 $\omega(t) = 90 - 45\ e^{-t}(t + 2)$ rad/s

6.5   a) $v(t) = (e^{-t} - e^{-2t})$ m/s

b) $v(t) = (\frac{1}{2} - e^{-t} + \frac{1}{2} e^{-2t})$ m/s

c) $v(t) = (-e^{-t} + 2\ e^{-2t})$ m/s

d) $v(t) = 2[e^{-t}-e^{-2t}]u_s(t)-[e^{-(t-2)}-e^{-2(t-2)}]u_s(t-2)$

6.7 $\omega(t) = (50 - 150\ e^{-2t} + 100\ e^{-3t})$ rad/s

$T_B(t) = (1\ 125 - 3\ 375\ e^{-2t} + 2\ 250\ e^{-3t})$ N·m

6.9 $v(t) = [20\ e^{-2t} - 20\ e^{-4t}]u_s(t)$

$-[20\ e^{-2(t-2)}-20\ e^{-4(t-2)}]u_s(t-2)$

$x(t) = [5 - 10\ e^{-2t} + 5\ e^{-4t}]u_s(t)$

$-[5 - 10\ e^{-2(t-2)}+5\ e^{-4(t-2)}]u_s(t-2)$

6.11 $i_R(t) = e^{-5\ 000t}\ (.01 - 50t)$, A

$e_L(t) = e^{-5\ 000t}\ (1 - 2\ 500t)$, V

6.13 24 kPa

6.15 $\theta(t) = [3.06t^2-6.89t+3.05-2.5\ e^{-t}-0.55\ e^{-8t}]$

6.17 $Q(t) = [125\ e^{-2\ 000t} - 110\ e^{-4\ 000t}](10)^{-6}$ coulombs

6.19 $P(t) = [2\ 000 - 8\ 000\ e^{-15t} + 6\ 000\ e^{-20t}]$

CHAPTER VII

7.1   a) $Z_{dp} = \dfrac{Z_1[(Z_2+Z_5)(Z_3+Z_4)+Z_3Z_4]+Z_2[Z_3Z_4+Z_5(Z_3+Z_4)]}{(Z_2+Z_5)(Z_3+Z_4)+Z_3Z_4}$

    b) $Z_{dp} = \dfrac{Z_1Z_5(Z_3+Z_4)+Z_1Z_2(Z_3+Z_4+Z_5)}{(Z_1+Z_2)(Z_3+Z_4+Z_5)+Z_5(Z_3+Z_4)}$

7.3   a) $Z_{dp} = \dfrac{RLC_1C_2s^3+LC_2s^2+R(C_1+C_2)s+1}{L_1C_1C_2s^3+(C_1+C_2)s}$

    b) $Y_{dp} = \dfrac{2.5\ s^2+3.5\ s+1}{5\ s^2+10\ s+2}$

7.5 $Y_{dp} = \dfrac{L_2C_1C_2s^3+R_2C_1C_2s^2+(C_1+C_2)s}{R_1L_2C_1C_2s^3+[R_1R_2C_1C_2+L_2C_2]s^2+[R_1C_1+R_1C_2+R_2C_2]s+1}$

7.7   a) $\dfrac{E_o}{E_i} = \dfrac{R_2(R_1C_1s+1)}{R_2(R_1C_1s+1)+R_1(R_2C_2s+1)}$

    b) $\dfrac{E_o}{E_i} = \dfrac{R_2C_1s+1}{(LC_2s^2+1)(R_2C_1s+1)+(R_2+Ls)R_1C_1C_2s^2+R_1(C_1+C_2)s}$

7.9 $\dfrac{V_3}{F} = \dfrac{bk/s}{(m_1s+b)[(b+m_2s+k/s)(m_3s+k/s)-k^2/s^2]+b[(-b)(m_3s+k/s)]}$

7.11 $\dfrac{E_o}{E_i} = \dfrac{2\ s+1}{5\ s+1}$

7.13 $\dfrac{Q_o}{Q_i} = \dfrac{1}{L_2 Cs^2 + R_2 Cs + 1}$

7.15 $p_o(t) = 3t/2 + 3(1-e^{-2t})/4$

7.17 $f(t) = 0.25\, e^{-6t} \sin 8t$

7.21 $e_o(t) = 0.5 - 0.173\, e^{-0.44t} + 0.673\, e^{-4.56t}$

CHAPTER VIII

8.1 $\omega(t) = \dfrac{T}{B[1+(\omega J/B)^2]^{\frac{1}{2}}} \cos(\omega t-\phi) - \dfrac{BT}{B^2+(\omega J)^2}\, e^{-Bt/J}$

   $M = \dfrac{1}{B[1+(\omega J/B)^2]^{\frac{1}{2}}}\ ,\quad \phi = \tan^{-1} \dfrac{\omega J}{B}$

8.3 $Q(t) = 35.4(10)^{-6} \cos(0.1t - 45°)\ \text{m}^3/\text{s}$

8.5 $\omega(t) = 0.52 \sin(5t - 87°)\ \text{rad/s}$

8.7 $2.58\ \text{N·s/m}$

8.9 at .4 Hz $b_{eq} = 3.98\ \text{N·s/m},\quad k_{eq} = 10.05\ \text{N/m}$

CHAPTER IX

9.1 $P = 2.84\ \text{W},\quad \Delta e = 1.68\ \text{V},\quad i = 1.68\ \text{A}$

9.3 $n = 5,\quad P = 0.5\ \text{W}$

9.5 $n = 1/2$, $P = 625$ W, $V = 111.9$ V

9.7 $\delta = b/(8\sqrt{km})$

9.9 $\delta = 2b/(3\sqrt{km})$

9.11   a) $f = 490.5$ N

       b) $m_e = 2.5$ kg

BIBLIOGRAPHY

1. Gardner, M.F. and Barnes, J.S. "Transients in Linear Systems," John Wiley and Sons, New York, 1942.
2. Guillemin, E.A. "Introductory Circuit Theory," John Wiley and Sons, New York, 1953.
3. Cheng, D.K. "Analysis of Linear Systems," Addison-Wesley, Reading, Massachusetts, 1959.
4. Koenig, H.E. and Blackwell, W.A. "Electromechanical System Theory," McGraw-Hill, New York, 1961.
5. Lytle, D.W. and Harman, W.W. "Electrical and Mechanical Networks," McGraw-Hill, New York, 1962.
6. Lynch, W.A. and Truxal, J.G. "Signals and Systems in Electrical Engineering," McGraw-Hill, New York, 1962.
7. Guillemin, E.A. "Theory of Linear Physical Systems," John Wiley and Sons, New York, 1963.
8. Seeley, S. "Dynamic Systems Analysis," Reinhold, New York, 1964.
9. Scott, R.E. "Elements of Linear Circuits," Addison-Wesley, Reading, Massachusetts, 1965.
10. Pearson, S.I. and Maler, G.J. "Introductory Circuit Analysis," John Wiley and Sons, New York, 1965.

11. Mac Farlane, A.J.G. "Engineering Systems Analysis,"
    Addison-Wesley, Reading, Massachusetts, 1965.

12. Sanford, R.S. "Physical Networks," Prentice Hall,
    Engleword Cliffs, New Jersey, 1965.

13. Manning, S.A. "Electrical Circuits," McGraw-Hill,
    New York, 1966.

14. Roe, P.H. "Networks and Systems," Addison-Wesley,
    Reading, Massachusetts, 1966.

15. Gemlich, D.K. and Hammond, S.B. "Electromechanical
    Systems," McGraw-Hill, New York, 1967.

16. Reswick, J.B. and Taft, C.K. "Introduction to Dynamic
    Systems," Prentice Hall, 1967.

17. Koenig, H.E., Tokad, Y. and Kesavan, H.K. "Analysis of
    Discrete Physical Systems," McGraw-Hill, New York,
    1967.

18. Shearer, J.S., Murphy, A.T. and Richardson, H.H.
    "Introduction to System Dynamics," Addison-Wesley,
    Reading, Massachusetts, 1967.

19. Cannon, R.H. "Dynamics of Physical Systems," McGraw-
    Hill, New York, 1967.

20. Blackwell, W.A. "Mathematical Modelling of Physical
    Networks," MacMillan, New York, 1968.

21. Williams, G. "An Introduction to Electrical Circuit
    Theory," MacMillan Ltd, London, 1973.

22. Gray, B.F. "Elementary Engineering Systems," Longman
    Group Ltd, London, 1974.

INDEX